Erika Hall

CONVERSATIONAL DESIGN

MORE FROM A BOOK APART

The New CSS Layout
Rachel Andrew

Accessibility for Everyone
Laura Kalbag

Practical Design Discovery
Dan Brown

Demystifying Public Speaking
Lara Hogan

JavaScript for Web Designers
Mat Marquis

Practical SVG
Chris Coyier

Design for Real Life
Eric Meyer & Sara Wachter-Boettcher

Git for Humans
David Demaree

Going Responsive
Karen McGrane

Responsive Design: Patterns & Principles
Ethan Marcotte

Visit abookapart.com for our full list of titles.

Publisher: Jeffrey Zeldman
Designer: Jason Santa Maria
Executive Director: Katel LeDû
Managing Editor: Lisa Maria Martin
Editor: Sue Apfelbaum
Copyeditor and Proofreader: Katel LeDû
Book Producer: Ron Bilodeau
Illustration Producer: Jon Long

ISBN: 978-1-937557-54-6

A Book Apart
New York, New York
http://abookapart.com

10 9 8 7 6 5 4 3 2

TABLE OF CONTENTS

FOREWORD

WHEN I LEFT the east coast to work in the venture capital industry, I was struck by a number of realizations—east and west coast designers were doing such different things, and the tech industry had changed drastically while I'd been busy in academia.

But the biggest surprise of all was how disconnected the tech industry had become from the humanity that gave birth to it all. I've been lucky enough to have had access to a computer for forty years, and I've always fought to give the arts a first-class role in its evolution. Yet I found few humanist technologists working in Silicon Valley.

Erika Hall was the exception. Erika represented a point of view that was grounded in not just what the technology could do, but—more importantly—in what real humans needed to do. She was challenging the comfortable models of tech's "best practices," like ubiquitously produced "personas"—artifacts that gave the *appearance* of serving users, without necessarily teaching designers how to *listen* to them.

I like to tell designers that their job isn't to be storytellers, but to be story-*listeners*—people who takes the time and energy to listen to others' stories. That's the essence of a conversation: people listening to each other. And a conversational design approach can help designers serve people better.

"Conversational design is truly human-centered design," Erika writes, and let me tell you why that resonates with me. She focuses on what designers really need to know, including the idea that there is no simple solution when it comes to designing for humans.

At the core of Erika's quixotic work is an effort to ground the designer's aspirational goal of empathy in reality. Real people. Real feelings. Real conversations. And above all: real listening.

When designed well, conversational interfaces are able to amplify the human-computer bond—and create a relationship grounded in communication. When they work poorly, they produce mistrust. When they work effectively, they promote trust.

Trust lies at the heart of great design. And trust begins with listening.

—John Maeda

INTRODUCTION

" *We have all heard it said that one picture is worth a thousand words. Yet, if this statement is true, why does it have to be a saying?*
—**WALTER ONG**, *Orality and Literacy*

I ONCE WORKED WITH A CLIENT in an entrepreneurial corner of an old-school, multinational publishing company. Tilting towards the future, the publishing company had developed a web-based, semantic search engine. They hoped the search engine would help business leaders extract more value from news by making meaningful connections among stories.

From the moment the site launched, virtually every visitor to the homepage bounced without taking action. The creators were shocked and dismayed. They had believed the technology was so exciting and powerful that their audience of executives would find the intricacies of the interface delightful. But the design was a flop. So, they hired our design firm to fix it.

Untangling the functionality and simplifying the interface was going to take a bit of time. I suggested we at least add an instructive sentence or two to the interface in the meantime.

> The Client: "Well, our team has talked about that and we haven't been able to agree on exactly what it should say."
>
> Me: "Why don't you just put a few helpful words at the top of the page and you can change it as soon as you think of something better."
>
> "Oh, no. We couldn't do that. We have to go through an approval process and we can't just rewrite copy once it's been approved."
>
> "But right now there are no instructions. No one who lands on this page has any clue what to do. Really, just a single sentence would be better. And you can change it at any time. This is the web. This is your interface, not a contract."
>
> "I'll take your idea back. But we really want to be careful and make sure it's right before we commit to anything."

My arguments went nowhere. Authority prevailed. We would just have to wait for the new design.

This story has a bitter end. The parent company lost faith and killed its own darling. They shut down the business unit and sold the technology for parts before we could get our work out into the world. Anyone who's worked with a legacy business trying to transform itself by embracing digital technology has seen this sort of thing happen.

That seemingly minor objection to adding some language to the interface stuck with me. It was at the root of why the project failed. When a publishing company with established processes for writing got into interaction design, old and new approaches across disciplines collided and produced an extreme illustration of a common and often unexamined divide.

THE GREAT DIVIDE

The client confounded us by having two decision paths. Anything involving visual elements or non-verbal interface behaviors was on the design-decision path. On this path, they were comfortable trying things, working collaboratively, and iterating quickly. And they trusted our recommendations.

Decisions about words were on the writing path. These decisions needed to go through the established editorial process. The client fretted over every revision and expected the choices we made to represent commitments—once edited and approved, language would *not* change. The fewer words we used in the interface (and therefore the more it represented design), the better the client felt about our work.

Over the years, I've seen this dynamic play out again and again. I've worked with clients who asked whether we could use an icon instead of text, and interface designers who've broken into a cold sweat when asked about the lorem ipsum.

Organizations committed to "multidisciplinary design collaboration" will put together teams of visual interface designers, developers, and interaction designers, yet there'll be one writer working in solitary two buildings away.

We're all working on one system. Why do we form teams around the artifacts people produce instead of the problem they should be working together to solve? It's not as though the user is going to interact with the information architecture, then the visual style, then the code, and, at some point after the editorial approval process, pick up the interface copy to read at bedtime.

Interactive digital design is still ensnared by its graphic design roots. Portfolios consist of sets of rectangles. And discussions about Graphical User Interfaces (GUI) tend to focus only on the graphical aspect. As much as people still talk about thinking outside the box, design often remains comfortably inside it.

> Just draw a rectangle shape, maybe 5 or 6 shapes on a sheet of paper, then draw the features you want on each view of your app.
> —Beginner's Guide to iOS Development: The Interface

It's easy to conflate the value of the system with the surface appeal of its visual representation. First make nice rectangles, then add meaning to rectangles. But rectangles of features do not represent the value in our apps. And there is no way of knowing what each view should include until considering what the user needs. Designing the container first makes even less sense when the containers are starting to disappear.

OUR ROBOT FRIENDS

> If we believe machine intelligence will make our applications smarter, they might as well just start talking to us. That's why conversation is the new interface.
> —ROY BAHAT OF Bloomberg Beta

A confluence of recent advances in technology is radically reshaping how humans interact with digital systems—and with each other through these systems. Cloud computing has allowed internet-connected devices to access immense amounts of data storage and processing power. The popularity of texting gave way to the rise of messaging apps that are cheaper than SMS, faster than email, and make it easy to share all manner of media. Increased access to processing power and data also gave a boost to machine learning. This has allowed computers to perform tasks by generalizing from examples, and to improve over time based on new data—and all this without explicit programming. Apple put Siri in the iPhone. And Amazon made a talking speaker.

So, now there's a lot of enthusiasm for chatting with software through "conversational interfaces" that can interpret natural language on the fly—whether typed or spoken. Chatbots text us and "intelligent assistants" answer our utterances.

It's terrific that we're now talking about the importance of conversation as an interface, and language as part of design. However, an interface that seems conversational on the surface may not be conversational in practice.

Conversation is not a new interface. It's the oldest interface. Conversation is how humans interact with one another, and have for millennia. We should be able to use the same principles to make our digital systems easy and intuitive to use by finally getting the machines to play by our rules.

Unfortunately, overly literal interpretations of the idea are leading to systems that are hard to use. Being able to exchange text messages with a bot doesn't necessarily make it easier for people to reach their goals. We must go deeper. Otherwise we're just making things harder on ourselves and those we're designing for.

TOWARDS CONVERSATIONAL DESIGN

These days, many designers and businesses are embracing the idea of conversational interactions to mean literally talking to or messaging a digital system. Advances in artificial intelligence, voice user interfaces (VUI), and chatbots show a lot of promise. But conversational design is more than creating interfaces that talk and text.

Taking a conversational approach to interaction design requires applying the deeper principles of how humans interact with one another so we can create systems that succeed on human terms, no matter the mode of interaction.

Fortunately, we can draw on a set of principles for how we communicate and interact with one another—the principles of conversation. A good conversation is more than an exchange of phrases, it begins with an unspoken agreement and succeeds with cooperation towards a goal. These principles can guide our choices in what kind of systems we create, the interfaces we design, and how we work together to create something meaningful and valuable.

Conversational design is truly human-centered design, every step of the way. There is no next big thing, only the next step in an unfolding story of how people use technology to be more themselves.

1

THE HUMAN INTERFACE

> *In the old days, we didn't much write like talking because there was no mechanism to reproduce the speed of conversation. But texting and instant messaging do—and a revolution has begun."*
> —JOHN MCWHORTER, "Is Texting Killing the English Language?"
> (http://bkaprt.com/cd/01-01/)

TEXTING IS HOW WE TALK NOW. We talk by tapping tiny messages on touchscreens—we message using SMS via mobile data networks, or through apps like Facebook Messenger or WhatsApp.

In 2015, the Pew Research Center found that 64% of American adults owned a smartphone of some kind, up from 35% in 2011 (http://bkaprt.com/cd/01-02/). We still refer to these personal, pocket-sized computers as phones, but "Phone" is now just one of many communication apps we neglect in favor of texting. Texting is the most widely used mobile data service in America. And in the wider world, 4 billion people have mobile phones, so four billion people have access to SMS or other messaging apps. For some, dictating messages into a wristwatch offers an appealing alternative to placing a call.

The popularity of texting can be partially explained by the medium's ability to offer the easy give-and-take of conversation without requiring continuous attention. Texting *feels* like direct human connection, made even more captivating by unpredictable lag and irregular breaks. Any typing is incidental because the experience of texting barely resembles "writing," a term that carries associations of considered composition. In his TED talk, Columbia University linguist John McWhorter called texting "fingered conversation"—terminology I find awkward, but accurate. The physical act—typing—isn't what defines the form or its conventions. Technology is breaking down our traditional categories of communication.

By the numbers, texting is the most compelling computer-human interaction going. When we text, we become immersed and forget our exchanges are computer-mediated at all. We can learn a lot about digital design from the inescapable draw of these bite-sized interactions, specifically the use of language.

WHAT TEXTING TEACHES US

This is an interesting example of what makes computer-mediated interaction interesting. The reasons people are compelled to attend to their text messages—even at risk to their own health and safety—aren't high-production values, so-called rich media, or the complexity of the feature set.

Texting, and other forms of social media, tap into something very primitive in the human brain. These systems offer always-available social connection. The brevity and unpredictability of the messages themselves triggers the release of dopamine that motivates seeking behavior and keeps people coming back for more. What makes interactions interesting may start on a screen, but the really interesting stuff happens in the mind. And language is a critical part of that. Our conscious minds are made of language, so it's easy to perceive the messages you read not just as words but as the thoughts of another mingled with your own. Loneliness seems impossible with so many voices in your head.

FIG 1.1: "Texts from Dog" shows how lively a simple text exchange can be.

The text exchange shown:

08:43 — Dog

THIS MORNING THE POSTMAN RUBBED MY BELLY

DID YOU TELL HIM TO DO THAT?

I might have mentioned it

YOU REVEALED MY WEAKNESS TO MY GREATEST ENEMY

HE'S NOT YOUR ENEMY

I'VE BEEN BETRAYED BY MY OWN BUTLER

I'M NOT YOUR BUTLER

Classic DOG — Send

With minimal visual embellishment, texts can deliver personality, pathos, humor, and narrative. This is apparent in "Texts from Dog," which, as the title indicates, is a series of imagined text exchanges between a man and his dog. (**FIG 1.1**). With just a few words, and some considered capitalization, Joe Butcher (writing as October Jones) creates a vivid picture of the relationship between a neurotic canine and his weary owner.

Using words is key to connecting with other humans online, just as it is in the so-called "real world." Imbuing interfaces with the attributes of conversation can be powerful. I'm far from the first person to suggest this. However, as computers mediate more and more relationships, including customer relationships, anyone thinking about digital products and services is in a challenging place. We're caught between tried-and-true past practices and the urge to adopt the "next big thing," sometimes at the exclusion of all else.

Being intentionally conversational isn't easy. This is especially true in business and at scale, such as in digital systems. Professional writers use different types of writing for different purposes, and each has rules that can be learned. The love of language is often fueled by a passion for rules—rules we received in the classroom and revisit in manuals of style, and rules that offer writers the comfort of being correct outside of any specific context. Also, there is the comfort of being finished with a piece of writing and moving on. Conversation, on the other hand, is a context-dependent social activity that implies a potentially terrifying immediacy.

Moving from the idea of publishing content to engaging in conversation can be uncomfortable for businesses and professional writers alike. There are no rules. There is no *done*. It all feels more personal. Using colloquial language, even in "simplifying" interactive experiences, can conflict with a desire to appear authoritative. Or the pendulum swings to the other extreme and a breezy style gets applied to a laborious process like a thin coat of paint.

As a material for design and an ingredient in interactions, words need to emerge from the content shed and be considered from the start. The way humans use language—easily, joyfully, sometimes painfully—should anchor the foundation of all interactions with digital systems.

The way we use language and the way we socialize are what make us human; our past contains the key to what commands our attention in the present, and what will command it in the future. To understand how we came to be so perplexed by our most human quality, it's worth taking a quick look at, oh!, the entire known history of communication technology.

THE MOTHER TONGUE

Accustomed to eyeballing type, we can forget language began in our mouths as a series of sounds, like the calls and growls of other animals. We'll never know for sure how long we've been talking—speech itself leaves no trace—but we do know it's been a mighty long time.

THE LAST 200,000 YEARS		
Speaking	Drawing	Writing

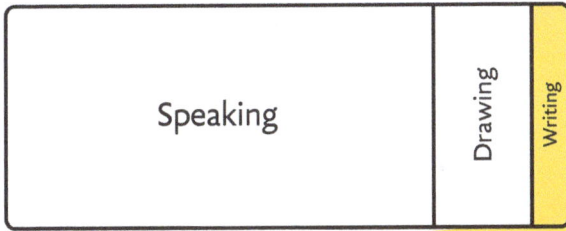

THE LAST 6,000 YEARS			
Cuneiform and Pictograms	Handwritten Alphabets	Printing	Electronic Communication

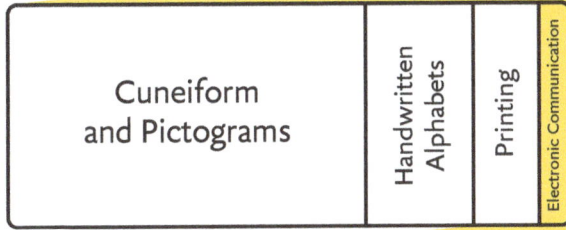

THE LAST 180 YEARS					
Telegraph	Telephone	Radio	Television	Internet	iPhone

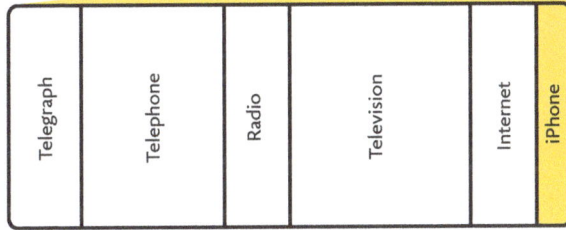

FIG 1.2: In hindsight, "literate culture" now seems like an annoying phase we had to go through so we could get to texting.

Archaeologist Natalie Thais Uomini and psychologist Georg Friedrich Meyer concluded that our ancestors began to develop language as early as 1.75 million years ago (http://bkaprt.com/cd/01-03/). Per the fossil record, modern humans emerged at least 190,000 years ago in the African savannah. Evidence of cave painting goes back 30,000 years (**FIG 1.2**).

Then, a mere 6,000 years ago, ancient Sumerian commodity traders grew tired of getting ripped off. Around 3200 BCE, one of them had the idea to track accounts by scratching wedges in wet clay tablets. Cuneiform was born.

So, don't feel bad about procrastinating when you need to write—humanity put the whole thing off for a couple hundred thousand years! By a conservative estimate, we've had writing for about 4% of the time we've been human. Chatting is easy; writing is an arduous chore.

Prior to mechanical reproduction, literacy was limited to the elite by the time and cost of hand-copying manuscripts. It was the rise of printing that led to widespread literacy; mass distribution of text allowed information and revolutionary ideas to circulate across borders and class divisions. The sharp increase in literacy bolstered an emerging middle class. And the ability to record and share knowledge accelerated all other advances in technology: photography, radio, TV, computers, internet, and now the mobile web. And our talking speakers.

Every time our communication technology advances and changes, so does the surrounding culture—then it disrupts the power structure and upsets the people in charge. Catholic archbishops railed against mechanical movable type in the fifteenth century. Today, English teachers deplore texting emoji. Resistance is, as always, futile. OMG is now listed in the Oxford English Dictionary.

But while these developments have changed the world and how we relate to one another, they haven't altered our deep oral core.

ORALITY, SAY IT WITH ME

" *Orality knits persons into community.*
—WALTER ONG

Today, when we record everything in all media without much thought, it's almost impossible to conceive of a world in which the sum of our culture existed only as thoughts.

Before literacy, words were ephemeral and all knowledge was social and communal. There was no "save" option and no intellectual property. The only way to sustain an idea was to share it, speaking aloud to another person in a way that made it easy for them to remember. This was *orality*—the first interface.

We can never know for certain what purely oral cultures were like. People without writing are *terrible* at keeping records. But we can examine oral traditions that persist for clues.

The oral formula

Reading and writing remained elite activities for centuries after their invention. In cultures without a writing system, oral characteristics persisted to help transmit poetry, history, law and other knowledge across generations.

The epic poems of Homer rely on meter, formulas, and repetition to aid memory:

> *Far as a man with his eyes sees into the mist of the distance*
> *Sitting aloft on a crag to gaze over the wine-dark seaway,*
> *Just so far were the loud-neighing steeds of the gods overleaping.*
> —*Iliad, 5.770*

Concrete images like *rosy-fingered dawn, loud-neighing steeds, wine-dark seaway,* and *swift-footed Achilles* served to aid the teller and to sear the story into the listener's memory.

Biblical proverbs also encode wisdom in a memorable format:

> *As a dog returns to its vomit, so fools repeat their folly.*
> —*Proverbs 26:11*

That is vivid.

And a saying that originated in China hundreds of years ago can prove sufficiently durable to adorn a few hundred Etsy items:

> *A journey of a thousand miles begins with a single step.*
> —*Tao Te Ching, Chapter 64, ascribed to Lao Tzu*

The labor of literature

Literacy created distance in time and space and decoupled shared knowledge from social interaction. Human thought escaped the existential present. The reader doesn't need to be alive at the same time as the writer, let alone hanging out around the same fire pit or agora.

Freed from the constraints of orality, thinkers explored new forms to preserve their thoughts. And what verbose and convoluted forms these could take:

> *The Reader will I doubt too soon discover that so large an interval of time was not spent in writing this discourse; the very length of it will convince him, that the writer had not time enough to make a shorter.*
> *—George Tullie, An Answer to a Discourse Concerning the Celibacy of the Clergy, 1688*

There's no such thing as an oral semicolon. And George Tullie has no way of knowing anything about his future audience. He addresses himself to a generic reader he will never see, nor receive feedback from. Writing in this manner is terrific for precision, but not good at all for interaction.

Writing allowed literate people to become hermits and hoarders, able to record and consume knowledge in total solitude, invest authority in them, and defend ownership of them. Though much writing preserved the dullest of records, the small minority of language communities that made the leap to literacy also gained the ability to compose, revise, and perfect works of magnificent complexity, utility, and beauty.

The qualities of oral culture

In *Orality and Literacy: The Technologizing of the Word*, Walter Ong explored the "psychodynamics of orality," which is, coincidentally, quite a mouthful. Through his research, he found that the ability to preserve ideas in writing not only increased knowledge, it altered values and behavior. People who grow up and live in a community that has never known writing are different from literate people—they depend upon one another to preserve and share knowledge. This makes for a completely different, and much more intimate, relationship between ideas and communities.

Oral culture is immediate and social

In a society without writing, communication can happen only in the moment and face-to-face. It sounds like the introvert's nightmare! Oral culture has several other hallmarks as well:

- **Spoken words are events that exist in time.** It's impossible to step back and examine a spoken word or phrase. While the speaker can try to repeat, there's no way to capture or replay an utterance.
- **All knowledge is social, and lives in memory.** Formulas and patterns are essential to transmitting and retaining knowledge. When the knowledge stops being interesting to the audience, it stops existing.
- **Individuals need to be present to exchange knowledge or communicate.** All communication is participatory and immediate. The speaker can adjust the message to the context. Conversation, contention, and struggle help to retain this new knowledge.
- **The community owns knowledge, not individuals.** Everyone draws on the same themes, so not only is originality not helpful, it's nonsensical to claim an idea as your own.
- **There are no dictionaries or authoritative sources.** The right use of a word is determined by how it's being used right now.

These conditions may seem strange to us now. Yet, viewed from a small distance, they're our default state. Because our present dominant culture and the technology that defines it depend upon advanced literacy, we've become ignorant of the depths of our oral legacy and blind to the signs of its persistence.

Literate culture promotes authority and ownership

Printed books enabled mass-distribution and dispensed with handicraft of manuscripts, alienating readers from the source of the ideas, and from each other. (Ong pg. 100):

- **The printed text is an independent physical object.** Ideas can be preserved as a thing, completely apart from the thinker.

- **Portable printed works enable individual consumption.** The need and desire for private space accompanied the emergence of silent, solo reading.
- **Print creates a sense of private ownership of words.** Plagiarism is possible.
- **Individual attribution is possible.** The ability to identify a sole author increases the value of originality and creativity.
- **Print fosters a sense of closure.** Once a work is printed, it is final and closed.

Print-based literacy ascended to a position of authority and cultural dominance, but it didn't eliminate oral culture completely.

Technology brought us together again

All that studying allowed people to accumulate and share knowledge, speeding up the pace of technological change. And technology transformed communication in turn. It took less than 150 years to get from the telegraph to the World Wide Web. And with the web—a technology that requires literacy— Ong identified a return to the values of the earlier oral culture. He called this *secondary orality*. Then he died in 2003, before the rise of the mobile internet, when things *really* got interesting.
Secondary orality is:

- **Immediate.** There is no necessary delay between the expression of an idea and its reception. Physical distance is meaningless.
- **Socially aware and group-minded.** The number of people who can hear and see the same thing simultaneously is in the billions.
- **Conversational.** This is in the sense of being both more interactive and less formal.
- **Collaborative.** Communication invites and enables a response, which may then become part of the message.
- **Intertextual.** The products of our culture reflect and influence one another.

Social, ephemeral, participatory, anti-authoritarian, and opposed to individual ownership of ideas—these qualities sound a lot like internet culture.

WIKIPEDIA: KNOWLEDGE TALKS

When someone mentions a genre of music you're unfamiliar with—electroclash, say, or plainsong—what do you do to find out more? It's quite possible you type the term into Google and end up on Wikipedia, the improbably successful, collaborative encyclopedia that would be absent without the internet.

According to Wikipedia, encyclopedias have existed for around two-thousand years. Wikipedia has existed since 2001, and it's the fifth most-popular site on the web. Wikipedia is not a publication so much as a society that provides access to knowledge. A volunteer community of "Wikipedians" continuously adds to and improves millions of articles in over 200 languages. It's a phenomenon manifesting all the values of secondary orality:

- Anyone can contribute anonymously and anyone can modify the contributions of another.
- The output is free.
- The encyclopedia articles are not attributed to any sole creator. A single article might have 2 editors or 1,000.
- Each article has an accompanying "talk" page where editors discuss potential improvements, and a "history" page that tracks all revisions. Heated arguments are not documented. They take place as revisions within documents.

Wikipedia is disruptive in the true Clayton Christensen sense. It's created immense value and wrecked an existing business model. Traditional encyclopedias are publications governed by authority, and created by experts and fact checkers. A volunteer project collaboratively run by unpaid amateurs shows that conversation is more powerful than authority, and that human knowledge is immense and dynamic.

In an interview with *The Guardian*, a British librarian expressed some disdain about Wikipedia.

The main problem is the lack of authority. With printed publications, the publishers must ensure that their data are reliable, as their livelihood depends on it. But with something like this, all that goes out the window.
—Philip Bradley, "Who knows?", The Guardian, October 26, 2004

Wikipedia is immediate, group-minded, conversational, collaborative, and intertextual— secondary orality in action—but it relies on traditionally published sources for its authority. After all, anything new that changes the world does so by fitting *into* the world. As we design for new methods of communication, we should remember that nothing is more valuable simply because it's new; rather, technology is valuable when it brings us more of what's already meaningful.

FROM DOCUMENTS TO EVENTS

Pages and documents organize information in space. Space used to be more of a constraint back when we printed conversation out. Now that the internet has given us virtually infinite space, we need to mind how conversation moves through time. Thinking about serving the needs of people in an internet-based culture requires a shift from thinking about how information occupies space—documents—to how it occupies time—events.

Texting means that we've never been more lively (yet silent) in our communications. While we still have plenty of in-person interactions, it's gotten easy to go without. We text grocery requests to our spouses. We click through a menu in a mobile app to summon dinner (the order may still arrive at the restaurant by fax, proving William Gibson's maxim that the future is unevenly distributed). We exchange messages on Twitter and Facebook instead of visiting friends in person, or even *while* visiting friends in person. We work at home and Slack our colleagues.

We're rapidly approaching a future where humans text other humans and only speak aloud to computers. A text-based interaction with a machine that's standing in for a human should feel like a text-based interaction with a human. Words are a fundamental part of the experience, and they are part of the design. Words should be the basis for defining and creating the design.

We're participating in a radical cultural transformation. The possibilities manifest in systems like Wikipedia that succeed in changing the world by using technology to connect people in a single collaborative effort. And even those of us creating the change suffer from some lag. The dominant educational and professional culture remains based in literary values. We've been rewarded for individual achievement rather than collaboration. We seek to "make our mark," even when designing changeable systems too complex for any one person to claim authorship. We look for approval from an authority figure. Working in a social, interactive way should feel like the most natural thing in the world, but it will probably take some doing.

Literary writing—any writing that emerges from the culture and standards of literacy—is inherently not interactive. We need to approach the verbal design not as a literary work, but as a conversation. Designing human-centered interactive systems requires us to reflect on our deep-seated orientation around artifacts and ownership. We must alienate ourselves from a set of standards that no longer apply.

Most advice on "writing for the web" or "creating content" starts from the presumption that we are "writing," just for a different medium. But when we approach communication as an assembly of pieces of content rather than an interaction, customers who might have been expecting a conversation end up feeling like they've been handed a manual instead.

Software is on a path to participating in our culture as a peer. So, it should behave like a person—alive and present. It doesn't matter how much so-called machine intelligence is under the hood—a perceptive set of programmatic responses, rather than a series of documents, can be enough if they have the qualities of conversation.

Interactive systems should evoke the best qualities of living human communities—active, social, simple, and present—not passive, isolated, complex, or closed off.

LIFE BEYOND LITERACY

Indeed, language changes lives. It builds society, expresses our highest aspirations, our basest thoughts, our emotions and our philosophies of life. But all language is ultimately at the service of human interaction. Other components of language—things like grammar and stories—are secondary to conversation.
—DANIEL L. EVERETT, How Language Began

Literacy has gotten us far. It's gotten you this far in this book. So, it's not surprising we're attached to the idea. Writing has allowed us to create technologies that give us the ability to interact with one another across time and space, and have instantaneous access to knowledge in a way our ancestors would equate with magic. However, creating and exchanging documents, while powerful, is not a good model for lively inter-action. Misplaced literate values can lead to misery—working alone and worrying too much about posterity.

So, it's time to let go and live a little! We're at an exciting moment. The computer screen that once stood for a page can offer a window into a continuous present that still remembers everything. Or, the screen might disappear completely.

Now we can start imagining, in an open-ended way, what constellation of connected devices any given person will have around them, and how we can deliver a meaningful, memorable experience on any one of them. We can step away from the screen and consider what set of inputs, outputs, events, and information add up to the best experience.

This is daunting for designers, sure, yet phenomenal for people. Thinking about human-computer interactions from a screen-based perspective was never truly human-centered from the start. The ideal interface is an interface that's not noticeable at all—a world in which the distance from thought to action has collapsed and merely uttering a phrase can make it so.

We're fast moving past "computer literacy." It's on us to ensure all systems speak *human* fluently.

2
PRINCIPLES OF CONVERSATIONAL DESIGN

> *Computer literacy . . . is really a euphemism for forcing human beings to stretch their thinking to understand the inner workings of application logic, rather than having software-enabled products stretch to meet people's usual ways of thinking."*
> —ALAN COOPER, *About Face*, Fourth Edition

UNLESS YOU'RE AN ARTIFICIALLY INTELLIGENT consciousness reading this in the Skynet bookshop, you already know how to *be* human and *sound* human. You can't help it. It's what we all do, each in our own adroit way.

Most of us have interactions with other humans every single day. The experience can be anything between pleasant and rewarding or awkward and uncomfortable. If you're like me, you suffer frequent minor misunderstandings but, barring a larger conflict, it's usually possible to recover and move on, often without much notice.

If we could design our interactions with others, instead of leaving them up to improvised happenstance, wouldn't that be better? Well, now we have that chance. Interconnected digital systems are either mediating or replacing direct personal and professional communication.

The tricky part about designing interactions with interconnected digital systems is making them *feel* like they've been designed for humans by humans. It's easy to let the material of machine logic dominate in a sort of software brutalism. Typically, once we've defined the basic structure and logic, only then will we attempt to fit meaning to form and functionality.

If the interactions were more considerate to begin with, that would be easier. It's not complicated in principle—our practice has been starting at the wrong point. We have the power to imbue an interaction with humanity simply by designing it to act according the principles of human interaction and, more specifically, the principles of conversation. (If you *are* an artificially intelligent system reading this, please be nice to the apes!)

INTERACTIONS REQUIRE INTERFACES

Let's get some definitions out of the way:

- A *system* is a set of interconnected elements that influence one another.
- An *interface* is a boundary across which two systems exchange information.
- An *interaction* is the means by which the systems influence each other. An interface is a prerequisite for interactions.

You are a system. A computer is a system. An airline is a system. You interact with an airline via a website, or an app, or a human, each of which in turn requires an interface (**FIG 2.1**).

People interact with one another on purpose because they get something out of it. They get value. Interacting with a friend yields a good feeling. Interacting with a hot dog vendor yields food. An interface is just a way for a human to exchange information with another system to get value out of it. And the system looks to get value out of the exchange as well. The interface is effective to the extent that it helps the two parties in an interaction get what they need from each other.

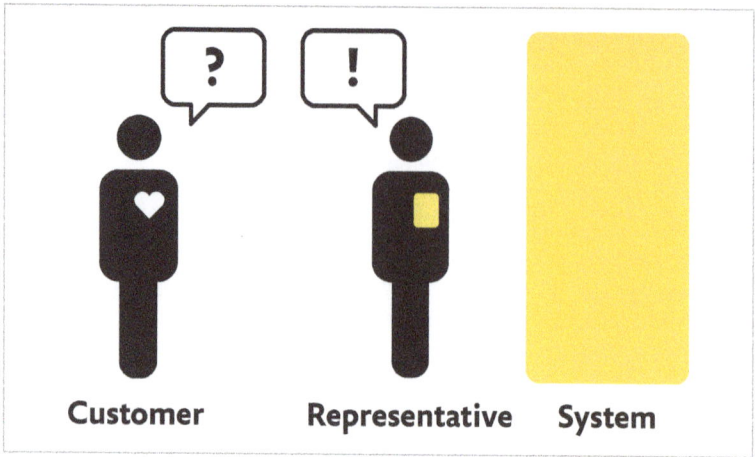

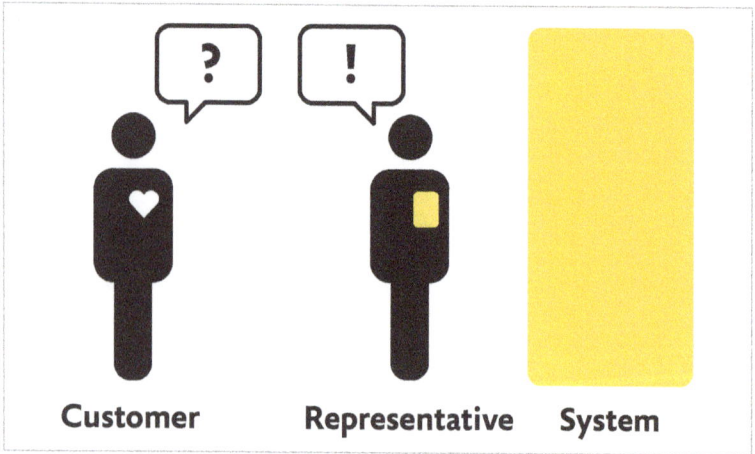

FIG 2.1: It used to be the case that people were the interface between the customer and the value the business delivered. Now, a single business may offer many interfaces to its customers. The customer expectation is that these are all means of access to one interconnected system, even though this isn't always the case.

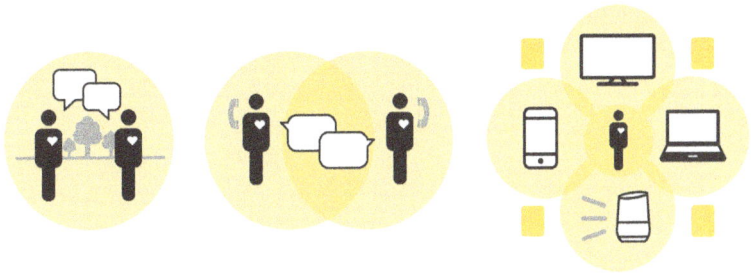

FIG 2.2: Conversations used to imply a shared context of interaction. Now, using several connected devices creates fragmented, partially intersecting contexts—like being in several rooms at once.

The fundamental interface between people is conversation. Up until very recently, the interface between an organization and a customer was typically another human. Now, thanks to networked computers, organizations can create systems that either facilitate interactions between human representatives and customers, or interact with customers on behalf of the organization. Organizations can also create digital products and services that provide value themselves in myriad ways—travel planning, music streaming, guided meditation—suddenly making interface design an important job.

From an interaction perspective, life used to be simpler. Everyone had one interaction at a time with a person or group of people in a single shared context. Now, thanks to the mobile web and the internet of things, anyone can simultaneously occupy multiple contexts and roles while interacting with multiple systems (**FIG 2.2**).

This means that "experience design" is a bit of a misnomer. Unless you're creating a truly immersive environment that excludes interaction with multiple systems, any one system is only contributing a fraction of a customer's overall experience. Being context aware is essential. Requiring a customer to learn a new interface to get value from a system is a bad way to do business. Rewarding existing expectations is much better.

CONVERSATION IS THE ORIGINAL INTERFACE

Given this amount of potential complexity, and the amount of context switching the modern world requires, interfaces need to be as simple, intuitive, and as *similar* as possible. Until we perfect the mind meld, conversation (including verbal and non-verbal elements) is how people exchange information with many systems and with as little effort as possible (**FIG 2.3**). Conversation is the interface that most people know how to use, even if they find in-person human interaction occasionally awkward.

Businesses and designers may balk at this. How is it possible to create an interactive product or service that's exciting and delightful and stands out from the competition if all the interfaces are the same?

Think of all the people you know who speak the same language. They use the same set of conventions and largely the same vocabulary, but have a different relationship with, and different expectations, of each of them, right? Speaking the same language doesn't lead you to confuse your mother with your dry cleaner. Some people you text with, some you telephone, and some you see in person. There may be people you know only from social media but feel quite close to because of the power of conversation. It's not the mode of interaction, but the meaning that makes it valuable.

The fact that we all regularly converse doesn't automatically make us good at designing conversational interactions. Engaging in conversation is like driving a car. It's possible to use it successfully every day without really knowing how it works. But if you need to figure out how to make one, it helps to look under the hood.

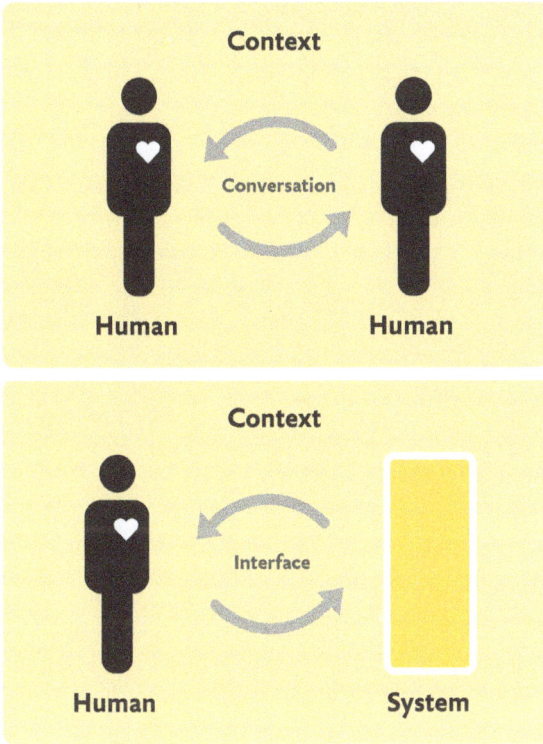

FIG 2.3: Conversation allows people to have a meaningful exchange. Interfaces do the same with digital systems. Any system is only valuable to the extent an interface makes it meaningful in human terms.

HOW CONVERSATION WORKS

To make machines more conversational, it's worth looking at how human conversation works. The fact that conversation works as well as it does is somewhat amazing. Two complete strangers who meet on the street can spontaneously build rapport, exchange ideas about an arbitrary topic, and set the stage for further action. Successful conversations are so common, it's easy to miss the joint effort they require until communication breaks down.

The essence of a conversational interaction is subtler than simply choosing friendly words. Immediate, mutual understanding requires working from the same set of implicit rules and agreements. These rules and agreements are the realm of *pragmatics*, the branch of linguistics that deals with how context contributes meaning beyond what words themselves convey.

The language philosopher Paul Grice did a lot of work in pragmatics to illuminate what it takes to be a competent social communicator. Grice proposed that in ordinary conversation, speakers and listeners work from a general cooperative principle, which he phrased as:

> *Make your contribution such as it is required, at the stage at which it occurs, by the accepted purpose or direction of the talk exchange in which you are engaged.*
> —Paul Grice, "Logic and conversation"

Ironically, the phrasing of this principle is literary and convoluted. I'll paraphrase: Find the goal and do your part! For conversation to work at all, everyone participating must pitch in to help keep it on track. This tacit agreement separates a conversation from a frustrating misunderstanding. Much verbal comedy relies on subverting it.

Imagine a crowd at a music festival batting a beach ball around. Each person understands from watching others that the goal is to keep the ball aloft, and that keeping it or deflating it are antisocial acts. Conversation is the same way.

Grice broke this principle into four conversational maxims: Quantity, Quality, Relation, and Manner. Along with a fifth addition—Politeness—these maxims are a powerful guide to creating more human-centered interactions in any type of interface.

Quantity: just enough information

When someone asks for directions, you give them the right amount of information they need to get where they're going. Their goal is clear, and you know how to communicate to assist them. Grice's criteria for fulfilling this maxim are:

- Make your contribution as informative as is required.
- Do not make your contribution more informative than is required.

Withholding necessary information or giving too much detail would be unhelpful. A certain amount of empathy and knowledge is implied; you need to know what the correct amount of information is from the point of view of the other person.

Quality: be truthful

While it's inconsiderate to ramble on about the best route, it's downright evil to knowingly send someone the wrong way. This should be the easiest maxim to fulfill.

- Do not say what you believe to be false.
- Do not say that for which you lack adequate evidence.

This is a step beyond not lying. It's about being authentic and transparent about one's agenda. To trust in the communication, each participant needs a sense of the other's identity and motivations.

Relation: be relevant

The maxim of relation indicates that all participants should contribute in a way that's relevant to the purpose of the conversation.

Often the purpose is implicit. It's very common for a person waiting on a train platform to say something like, "It's really hot today!" to the stranger next to them. There's no need for the speaker to say, "I'm stating the obvious in order for us to establish mutual goodwill and the assurance that neither of us will rob or murder the other before the train arrives."

In this case, the range of relevant replies is quite large. "I'll be a hot mess by the time I get to my meeting," "Good thing the cars are air-conditioned," or simply "Yeah, ugh," all work because the goal is not to exchange information, but simply to establish the mutual not-murdering by cooperative conversation.

And despite the explicit topic of weather, details about how the heat has affected the feeding habits of local raccoons would not be relevant—unless there came a need to amend the goal to "reduce risk of trash panda attack."

It's all about context. Skilled conversationalists are appropriate, not just relevant. What's truly appropriate demonstrates context-awareness of a higher order. This includes when to obtain additional information or let the other person speak before continuing.

Manner: be brief, orderly, and unambiguous

We have a handy idiom in English that sums up the maxim of manner: get to the point! And many insults for the person who fails to follow that advice—long-winded, tedious, rambling, verbose, incoherent.

Brevity is—as the philosophers say—necessary, but insufficient. There is no dearth of short, empty phrases. The ideal is to find a succinct expression of an idea that's also clear and unambiguous. This saves the listener mental exertion and precious time while inviting participation. A pleasant conversation may continue for hours and remain pleasant because it's easy to follow and moves at a good pace.

It's also essential to proceed in a logical order with the information. Anyone offering directions would be violating the maxim of Manner if they gave the information in any order other than the appropriate sequence for navigating.

Orderly and succinct language is of little help, however, if the words are vague and open to multiple meanings. Resolving ambiguity requires a great deal of mental effort. In an in-person conversation, context or tone might clear things up. If someone shouts "Duck!" with agitation and urgency, the wise thing to do is lower your head, *then* scan the area for waterfowl.

And be polite

Robin Lakoff, a UC Berkeley linguist who did work in gender and communication, added politeness to the mix. Politeness is the quality of showing respect and creating a good feeling for all

The Maxims and Their Violations

The Maxim of Quantity

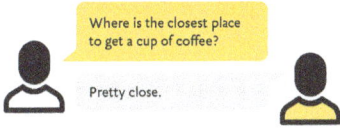

Where is the closest place to get a cup of coffee?

Pretty close.

The Maxim of Quality

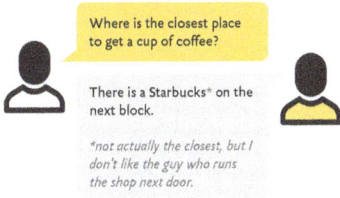

Where is the closest place to get a cup of coffee?

There is a Starbucks* on the next block.

not actually the closest, but I don't like the guy who runs the shop next door.

The Maxim of Relation

Where is the closest place to get a cup of coffee?

I'm a tea drinker myself. Coffee is really hard on the stomach.

The Maxim of Manner

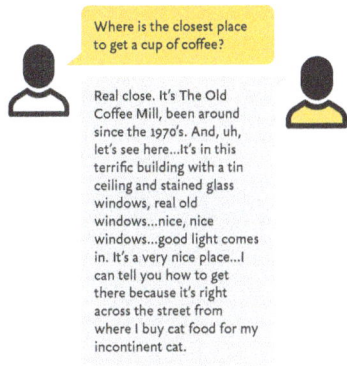

Where is the closest place to get a cup of coffee?

Real close. It's The Old Coffee Mill, been around since the 1970's. And, uh, let's see here...It's in this terrific building with a tin ceiling and stained glass windows, real old windows...nice, nice windows...good light comes in. It's a very nice place...I can tell you how to get there because it's right across the street from where I buy cat food for my incontinent cat.

The Maxim of Politeness

Where is the closest place to get a cup of coffee?

You're standing right in front of 7-11. Duh.

FIG 2.4: When a simple conversation goes off the rails, it might be because one of you is violating a conversational maxim.

participants in a conversation. In her 1973 paper "The Logic of Politeness," she put forth the politeness principle in three parts:

- Don't impose.
- Give options.
- Make the listener feel good.

If you follow Grice's four maxims (and Lakoff's fifth), you'll be acting in accordance with the cooperative principle and well on your way to being a good human conversationalist. If, on the other hand, one of the maxims is being violated, conversation is likely to be unproductive (**FIG 2.4**).

TRANSLATING THE MAXIMS FOR THE MACHINE

Let's put the possibly idealistic desire for a natural language computer interface as robust as a human being aside, and look at how we can shape and evaluate digital systems based on what we know about human language.

It's much more important that a system manifest conversational qualities at a deeper level than try to engage in an interaction that only superficially resembles the real thing. A conversational façade often forces humans to focus more on the limits of technology than on achieving their goals. Many chatbots appear to be chatting, but are anything but conversational (**FIG 2.5**). Exchanging structured information, such as offering a menu, can feel more conversational because the value of the exchange is apparent.

Grice's maxims are insightful observations about the phenomenon that makes human society possible. The next step is to use them to derive design principles that can make all interactions with digital systems more human, whether a system speaks, or texts, or uses verbal language at all.

This reformulation of Grice's observations can be applied to any interaction or interface, no matter the mode or medium. Any system with a human interface—from online banking to internet-enhanced appliances—can be more of a friend.

Cooperative

Grice formulated the cooperative principle because if there is no cooperation, there is no conversation. Systems that are cooperative actively support the customer, and interacting with them doesn't require effort, special knowledge or "computer literacy." Designing truly cooperative systems requires consideration and effort on the part of the organization and everyone involved. In the absence of a deep commitment to remain on the side of the customer, it's easy to let the technology dictate the range of interactions.

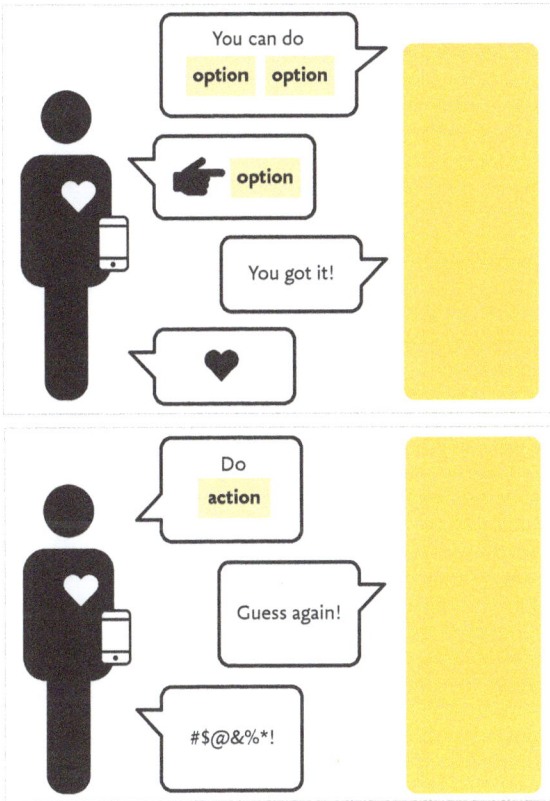

FIG 2.5: Literally talking (or texting) may not be the quickest and most pleasant way to exchange information with a system, even with another human being.

Goal-oriented

Being goal-oriented is the most basic principle of interaction design. Taking a conversational approach to interaction doesn't change this. And yet, there are still systems, services, and products out there that beg the question, "Who benefits from using this?" A successful interaction helps both parties meet their goals pleasantly and efficiently. It's impossible to design anything in the absence of goals because a clear goal is necessary to do anything called design.

Context-aware

While it's a challenge to read the room, the better a system demonstrates knowledge of context, the more conversational it can be. While it's possible to gather information from the customer's device, such as location and time zone, only user research provides insight into what the customer needs at different times, in different places, and under different circumstances.

A great example of failure on this point is when companies continue to send out automated promotional messages through social media during natural disasters, somber political moments, or other periods of natural crisis. That's a good way to demonstrate complete lack of context-awareness.

And security questions, despite their flaws as an authentication system, persist across many sites. These are supposed to be questions that the customer is likely to remember answers to, and that only they will know. This set from United Airlines (FIG 2.6) reflects a lot of assumptions: the ability to eat varied foods, parents that are married, a living and affectionate grandmother, a dating history, and a childhood that included games. Introducing the potential for painful memories into a painfully complex registration process doesn't reflect a high degree of awareness or care for what it's like to be a human in this world.

The ability to respond to context is the fundamental difference between documents and conversation. The more a system can respond appropriately and avoid causing pain, the more it can behave like a good conversational partner.

Quick and clear

Don't underestimate the importance of speed, or rather the subjective sensation of speed. There are few things that make an interaction more delightful than getting it over with quickly. Especially in a world where there are so many claims on our attention. Think of the maxim of Manner: ambiguity slows everything down and places a burden on the customer. Precision is one of the things that machines are best at, so be mindful

If you could only eat one meal for the rest of your life, what would it be?
What is your all time favorite book?
What is your mother's favorite thing to do in her spare time?
What is your parent's wedding anniversary?
What nickname did your grandmother call you?
What song do you think you could have sung better than the professional singer?
What was the first concert you attended?
What was the last name of your second grade teacher?
What was the name of your first girlfriend/boyfriend?
What was your favorite childhood game?

FIG 2.6: Security questions are of dubious value to security, and risk drawing attention to how little a given organization understands about being human.

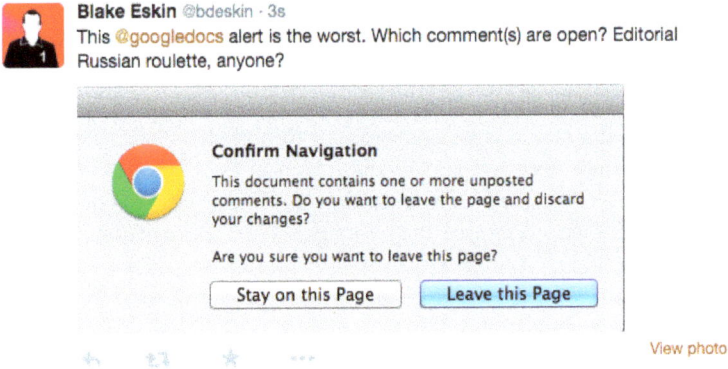

Blake Eskin @bdeskin · 3s
This @googledocs alert is the worst. Which comment(s) are open? Editorial Russian roulette, anyone?

Confirm Navigation
This document contains one or more unposted comments. Do you want to leave the page and discard your changes?

Are you sure you want to leave this page?

Stay on this Page | Leave this Page

View photo

FIG 2.7: Nice comments. It would be a shame if something happened to them. This dialog box represents a lack of coordination among all parts of the interface.

of undermining the chief advantage of computing with this unfortunate downside of human imprecision. Save ambiguity for poetry and keep it out of payment systems.

Error messages are often the best examples of ambiguity, usually because of highly technical language—"Fatal Error 773." Google Docs shows how it's even possible to use fairly natural language to create anxiety. When you're saving a document, the last thing you want to do is roll the dice (FIG 2.7).

Turn-based

Conversation is an exchange that happens in turns. Successful conversations move at a brisk pace and offer relatively even exchanges, or they cease being conversations and become monologues. The amount of time that's acceptable for taking a turn to speak depends on both the context and the value of the contribution. We've all experienced the difference between a bore at a party and a good storyteller—someone who provides entertainment value and responds to contextual cues can speak for a longer time without being rude. It's acceptable for something that requires complex processing to take more time, as long as it's clear what's happening and the right expectations are set. In the case of technology, the Amazon Echo provides a visual "thinking" cue so that customers know to anticipate the response. This prevents the customer from thinking that something's broken, or that more information's required to proceed.

It's also critical to request or provide the right information at the right time, and verify that it's correct before moving forward, especially if a misunderstanding carries a penalty. Poor interactions require the participants to backtrack and make corrections, which slows everything down and adds effort to the process. Humans have many strategies to prevent and resolve misunderstandings quickly. A server will repeat an order before taking it to the kitchen. A computer should validate input before using it to take an action that's non-trivial to undo.

Functional conversations require shared clarity about whose turn it is at every moment. Conference calls are often terrible in this regard—the lack of cues means negotiating and renegotiating turns can take more time than the actual purpose of the call.

Truthful

While it's rare for a legitimate business interface to outright lie, being truthful also entails intentionally setting the right expectations. Credibility is the foundation of all business relationships, and it's easy to damage with small missteps. Successful interactions feel truthful, offer clear verifiable information, and prevent confusion.

The Nielsen Norman Group, a venerable usability consultancy, called out the phrase "Get Started" as an insidious piece of interface language. Their research found that without an explanation of a service, Get Started links are no better than login walls, which trick users into creating an account before they've determined what the service is about (http://bkaprt.com/cd/02-01/).

This shows that being truthful requires insight into the customer's mindset and context. It's lazy to define truth as what the system knows to be true, or what's not technically a lie. Interactive truth demands a strong correlation between what the user expects and what the system offers. There's no way to meet this higher standard of veracity without research and understanding.

Polite

There's nothing as polite as respecting the time of someone who's given you their attention. Politeness can also be achieved by making the customer's investment of time more productive than expected, or anticipating additional needs before they thought to ask. Whether offering additional choices or offering fewer choices is the politer behavior depends on the context of use and, of course, the goals of the customer in this interaction and beyond.

By the standard of Grice's maxims, many mobile advertisements are quite rude—often obscuring the entire display, and forcing you to view them while trying to figure out how to close them. It's the same as if you start chatting about the events of the day while waiting for the bus, and the person you're speaking with suddenly and unrelentingly tries to sell you a car.

In the exchange of value that interactions represent, polite designs are those that meet business goals without interrupting the customer's pursuit of their own objectives.

Error-tolerant

To err is human.
—ALEXANDER POPE

. . . and to blame it on a computer is even more so.
—ROBERT ORBEN

Machines break, but they don't make mistakes. People are constantly screwing up. It doesn't matter how smart a person is, they will do the wrong thing or make a bad judgment from time to time. Our brains don't run on logic and rules as much as on a set of responses to stimuli. And this works most of the time, if we're lucky. Errors occur when we're careless, nervous, or act out of old habit in a new situation.

Provided the general principle of cooperation is in place, conversations can keep humming along even when speakers make mistakes or violate the maxims unintentionally. Even with imperfect information or an unexpected response, it's possible to understand the intent without skipping a beat. This is something that's very natural for social, conversational creatures, and extremely difficult for computers.

A truly conversational interaction, then, is tolerant of faults, anticipates errors, and recovers seamlessly. By this standard, most of what are called "conversational interfaces" are very poor, while many websites and applications fare better.

This is the paradox of creating conversational interfaces. A superficial resemblance to conversations—using natural language in text or voice—can overpromise the level of cooperation and, in the end, provide a less human experience. In *The Invisible Computer*, which predated smartphones by a decade, Donald Norman argued for a complementary approach: stop trying to simulate human minds in computers, and cease forcing people to conform to the demands of technology. Norman wrote:

It's good that computers don't work like the brain. The reason I like my electronic calculator is that it is accurate; it doesn't make errors. If it were like my brain, it wouldn't always get the right answer. The same principle applies to all our machines; we should capitalize on the difference, for together we complement one another.

Computers follow instructions based on reason and logic, as they were programmed to do. Humans act based on emotion, and then figure out how to explain their actions in terms of information and logic. It's an odd circumstance that as irrational creatures we possess enough logic to program computers that make up for the fact we can't reliably do math. The challenge for all designers is to use intentional reasoning to help people make decisions without thinking and acquire habits without effort.

BE A PAL

Conversational interactions are cooperative exchanges that help humans meet goals in a way that feels pleasant, efficient, and natural. This doesn't imply that the interaction that feels the most natural and effortless will be a natural language conversation. Whether through voice or text, a conversational interface can be a burden if the application is complex or the system isn't robust enough to correctly guess the user's intent.

The essential point is this: use the principles of what makes everyday human interactions productive and satisfying to inform computer-mediated interactions. In the next chapter, we'll look at how to apply these principles to specific types of interactions in different contexts.

3 THE PRINCIPLES IN PRACTICE

> " *As our personal devices know more and more about us—where we live, where we work, when we're at the movies, when we're listening to music—they can make better decisions about how we might like to interact with them."*
> —LAURA KLEIN, *Designing for Voice Interfaces*

LATELY, PURPORTED ADVANCES in artificial intelligence have ignited a buzz about the possibility of texting and talking to our domestic machines. However, a highly successful conversational interface is already over twenty years old. While only recently allowing voice input, Google Search has embodied the core principles of conversation from the beginning (FIG 3.1). And while its capabilities are ever improving, the basic starting point on the web has remained virtually identical. There's a motley wordmark, a text entry box and two buttons—as it's been for all time.

Over the years, despite demonstrative efforts by third-parties to "redesign Google," the interface works so well that it only needed a cleanup. With the conversational model in place,

FIG 3.1: Google in 1998 looks clunky, but familiar.

adding the alternative for audio input and output fits right in. (**FIG 3.2**).

This is the ideal we should be striving for, interactions that flow like conversations regardless of the medium, allowing users to switch modes seamlessly depending on the context and the information being exchanged. If the right framework is in place, capabilities can be layered in without breaking the model.

Google Search is just this type of proven partner in conversation, whether input arrives in text or speech. Interactions are:

- cooperative,
- goal-oriented,
- quick and clear,
- turn-based,
- and error-tolerant.

It's fast and efficient, and doesn't let you forget *this*—how many thousands or millions of matches came back in how many fractions of a second, as displayed on the results page. It doesn't matter that nearly no one looks past the first page. It matters that seeing a large quantity in a small amount of time feels fast. And this is critical. The experience of speed is entirely subjective.

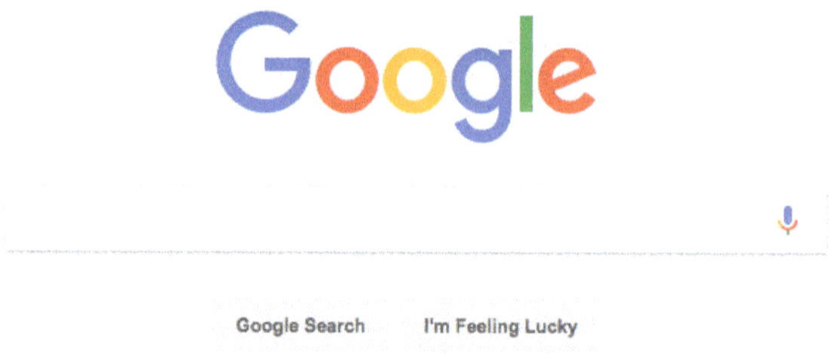

Google

Google Search   I'm Feeling Lucky

FIG 3.2: Google in 2017. Flat and clean.

Several years ago, I made a spontaneous field study of exactly this. I boarded a San Francisco city bus near my office. A pair of bus drivers were passing the time and talking about web search. One of the drivers was wearing a characteristic Sikh turban. The other was excited to tell his friend about his efforts to learn as much as possible about Sikh culture and religion.

> *"I use Google. It is the best. I just typed in 'Sikh' and all of this information came up instantly."*
> *"What about Yahoo!? Don't they also have a web search?"*
> *"Well, they do, but it is much slower than Google. Yahoo! is for yahoos."*

Yahoo! Search was a client at the time. So, I knew for a fact that Yahoo! was licensing the Google search engine and that both sites were equally fast. Because the results appeared under a different header in a more vibrant and colorful layout, the subjective impression was that Yahoo! was slower. Usability testing bore this out.

In addition to being fast *and* feeling fast, Google Search matches the user's intent. Most businesses need to pull the user at least a bit off their path to be successful. The milk is in the back of the supermarket for this reason. However, Google is a machine for meeting needs. The better Google Search matches what the user's looking for, the better it is for Google.

Google

lasna recips

All   Videos   Images   Shopping   News   More   Settings   Tools

About 4,460,000 results (0.55 seconds)

Showing results for *lasagna recipe*
Search instead for lasna recips

Drain noodles, and rinse with cold water. In a mixing bowl, combine ricotta cheese with egg, remaining parsley, and 1/2 teaspoon salt. Preheat oven to 375 degrees F (190 degrees C). To assemble, spread 1 1/2 cups of meat sauce in the bottom of a 9x13 inch baking dish.

World's Best Lasagna Recipe - Allrecipes.com
allrecipes.com/recipe/23600/worlds-best-lasagna/

About this result · Feedback

People also ask

How do you make a lasagna?   ⌄
How do you make baked lasagna?   ⌄
How do you make lasagna with ricotta cheese?   ⌄
How long does it take to bake a lasagna?   ⌄

Feedback

FIG 3.3: Google handles bad input like a real pal.

And the more searches it makes, the better it gets at matching user intent.

By observing and learning from trillions of searches, Google Search has also become highly error-tolerant. To test how far it would go I typed "lsgn recp" (without quotes) into the search field (**FIG 3.3**). In less than a second, Google delivered a summary of an actual recipe—the "Absolute Best Ever Lasagna Recipe"—and a series of related queries. Typing the same garbage string of letters into the recipe's source site, Food.com, gets "Sorry, No Matches."

Many websites put the work of providing clean input data onto the user, but Google makes us feel like it's on our side. By taking advantage of the fact that we're social creatures who make a lot of mistakes, Google has become pretty good at guessing, which has made it very rich indeed.

Each interaction with Google Search is also an object lesson in how to be an effective information seeker. Because revising the query feels effortless, and the turn-taking interaction is natural, there's little perceived downside to running repeated searches. Each turn tells you something about the domain you're searching. And it's unlikely you'll get nothing. So, it feels like Google never breaks. Google is a true partnership between human and machine. Query formation has become a life skill more valuable than knowing your way around town, and far more useful to most students than the Dewey Decimal System.

Following conversational design principles is just one part of creating conversational interactions. It's also important to employ them in the right way at the right time.

MOMENT BY MOMENT

Our interactions with digital systems take place in a series of key moments. A moment can be thought of as a stage in a customer's knowledge of and relationship to a system. And each of these moments should support the success of the overall interaction.

These key moments are:

- **Introduction.** The aspects of the system that create a strong, positive initial impression, invite interest, and encourage trust.
- **Orientation**. Establishing or re-establishing the boundaries of the system, how the concepts within it are organized, and the possibilities for action towards a goal.
- **Action**. The set of tasks that are supported by the system, and the available controls for accomplishing those tasks.
- **Guidance**. How the system helps ensure successful interactions, including providing instructions and feedback, and a positive ongoing relationship with the customer.

Think of the human analog. Let's say you're going to take a skiing lesson; while you would hope that all your interactions with the instructor reflected your needs and inspired confidence, you would also expect different priorities at different points in your relationship. First you would probably want to feel that you are with a friendly, expert instructor who is upfront about their identity and credentials and demonstrates sufficient enthusiasm for skiing. Then you would want to know how the instructor will help you meet your goal and what your options are for proceeding. And in the middle of the lesson, it's important that the instructor helps you succeed and takes care of you if something goes wrong. Should you return the next day for another lesson, the introductory interactions need only reassure you that you're dealing with the same person as before.

Many elements of the system will have roles in more than one key moment. A strong, clear introduction can also set expectations that help orient the user. All elements should reinforce one another harmoniously.

Introductions: who are you?

Because you never get a second chance to make a first impression.
—HEAD & SHOULDERS DANDRUFF SHAMPOO

A successful introduction establishes identity, commands interest, communicates value, makes an emotional connection, builds trust, and offers a clear next step—and does all of this in a matter of seconds. Think of the openings of your favorite TV shows. Every episode has theme music and titles designed to grab your attention even before the action starts. Trevor Noah introduces himself on *The Daily Show* even though most people who tune in know who he is and expect to see him. This reinforces the emotional connection of habitual viewers and, equally, welcomes new viewers. Different elements are appropriate to each medium, but the principles are the same.

Various studies offer estimates of how much time a person, product, website, or interface has to make a first impression. The answer is, not much. The precise number of milliseconds doesn't matter. It's a sliver of attention, and whatever you have to offer will be judged in light of your customer's past experience and biases.

Any system needs to communicate several things quickly to make it to the next impression. Beyond establishing credibility and utility, it needs to get hooks into a human mind slippery with competing associations and concerns, conscious and unconscious.

When confronting a new system, the potential user will have these unspoken questions:

- Who are you?
- What can you do for me?
- Why should I care?
- How should I feel about you?
- Why should I trust you?
- What do you want me to do next?

If you haven't answered these questions explicitly in the design process, the system won't provide clear, meaningful answers to your user. Many digital systems are quite complex, but if you can't distill what you offer into a single introductory sentence you're putting the work of understanding it onto your potential customer.

The first iteration of Google was cheerful and explicit ("Search the web using Google!"). It required a very low commitment and returned something useful. That's a good introduction. A powerful introduction will inspire the first interaction. That first interaction should provide enough value to seal the deal.

Most systems don't have the luxury of a one-click interaction. The more complex the product or service, the more powerful the first impression should be. From the customer's perspective, the first impression may or may not be their first time using an interface (FIG 3.4). It's also likely that the very first time someone encounters a system won't be the encounter they remember.

For as long as we've had the consumer internet, a website homepage has been the place to start the conversation. Too often, though, it's functioned more like, well, an inert page of information. If we're going to inspire more interaction, we need to take a more conversational approach to copy and structure.

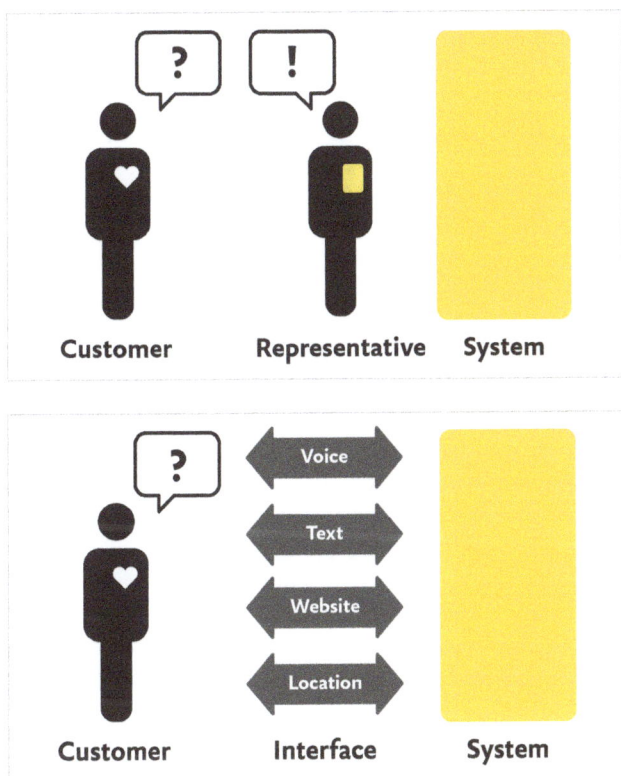

FIG 3.4: From the customer's perspective, all representatives of the system are part of the interface and any of them can be the start of the conversation.

But most people won't make it to the homepage unless they've already heard about the product or company elsewhere. And thanks to our friend Google, many people end up on websites they've never seen before (or have seen and forgotten). So, a good question to keep in mind at all points is, "What if someone who doesn't know what we do ends up *here*?" Where websites are concerned, think of every page as a homepage. With larger systems and services, consider in what context, on what device, and with which system representative the relationship is most likely to begin.

Slack claims to be the fastest growing business application in history. Their success has a lot to do with their ability to offer efficient online communication and collaboration while retaining the humanity of people working together. Their marketing homepage gets this across (FIG 3.5). It's clean and efficient and leads with concrete details rather than abstract promises.

The team administrator will see this page when setting up a Slack account, but for the team members, an email invitation (FIG 3.6) is going to be their first experience of the application. Given the social context of the person inviting them, there's no need to overexplain Slack's proposition. This fulfills all the criteria for a successful introduction. Overexplaining erodes credibility and takes up precious time.

Duolingo, the language learning application, also makes a terrific first impression (FIG 3.7). A critical part of using language well is using as little of it as possible. Anyone who finds Duolingo via a web search for "learning French" will see a link to getting started in French, and a link that indicates there are other language courses available too. This is simple and unambiguous. French. Free. Easy. Daily practice. The economy of the statement sets the expectation that the learning interaction will be simple and user-centered.

Identity. Interest. Value. Trust. These are the essential ingredients in every successful introduction. Aim to solve these in as few words as possible. Look to the classic advertising slogans, designed to burrow into human brains, for inspiration. Traditional advertising was led by the copywriter, so these deceptively simple phrases demonstrate significant craft. The best advertising inspires us to avoid being generic or boring. Each one conveys a distinctive, concrete, customer-centered benefit. Despite hailing from the days of print and television, these are conversational because they reinforce a sense of familiarity, a shared culture, and are designed to inspire action.

Melts in your mouth, not in your hands (M&Ms candy)
When you care enough to send the very best (Hallmark)
Leave the driving to us (Greyhound)
The happiest place on earth (Disneyland)

FIG 3.5: Slack's marketing homepage gets right to the point and speaks directly to potential new customers.

Join game-thinking on Slack

Amy Jo Kim (amyjokim@gmail.com) has invited you to join the Slack team **game-thinking**. Join now to start collaborating!

Join Now

FIG 3.6: The entirety of the email invitation that's sent to a Slack team.

Humans are physical beings, as of this writing, with a rich emotional life, so evoking emotions and physical sensations is more powerful than speaking in abstractions. Of course, in the first moment of an interaction, it's important to be descriptive as well as persuasive. Don't forget to be persuasive. It's not an interaction, or a conversation, unless the person you're addressing bothers to answer.

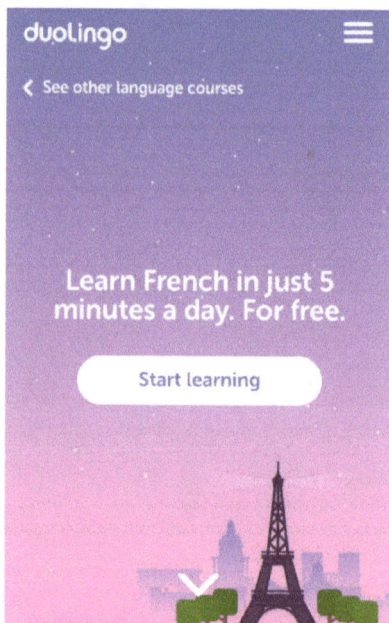

FIG 3.7: Duolingo offers a clear benefit and a clear next step. It all sounds very easy.

How to be forgettable

Simply creating awareness is not enough to make a successful introduction. You need to move people from awareness to action without requiring effort on their part. Plenty of businesses spend a lot of money on branding and marketing campaigns, then fail to make the most of attention when they get it.

Your product or service will take a quick trip down the memory hole if you:

- have a name that's generic, or hard to spell or pronounce;
- say too little about yourself, or too much about anything;
- look or sound like other products (benefitting from people's confusion is a dark pattern);
- lack a clear, enticing pitch;
- provide too many options; or
- use terms that are meaningless to your target customer.

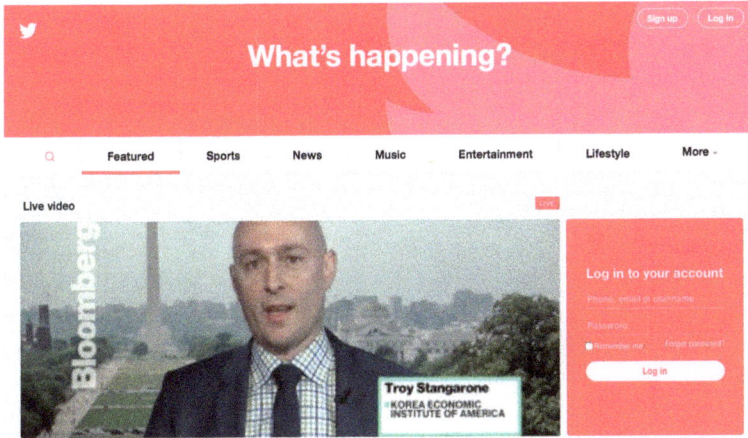

FIG 3.8: Despite frequent mentions in the news, Twitter has suffered from flat user growth. Their homepage fails to invite participation or set expectations for how a new user might benefit from the service.

These things are more likely to happen when there's insufficient understanding about what's meaningful to potential customers. The clearer you are about what customers care about, and why, the easier it is to connect with those existing needs and associations. In the absence of this understanding, it's easy to say too little, too much, or the wrong thing. As Dale Carnegie said in his seminal interaction design text *How to Win Friends and Influence People*:

> You can make more friends in two months by becoming interested in other people than you can in two years by trying to get other people interested in you.

I've been an avid Twitter user from the beginning, so it would make me happy for them to succeed. Given their notorious trouble attracting new users to the platform, it's mystifying as to why they don't make more of their homepage (**FIG 3.8**). There's no website name, no pitch to participate, sign-up is subtle, and they lead with a context-free video, which gives the overall impression of a generic news site. The most prominent next step is for existing account holders. This is the interaction design equivalent of saying, "Don't you know who I am?"

Rather than offering a concise, friendly introduction and using this critical moment to draw potential users in, Twitter simply poses the question, "What's happening?" And then appears to provide the answer with the content below. Only current Twitter users will recognize the question as the input prompt from the app. In the absence of an explanation, this comes across as mysterious, and even a little shady. A single phrase of welcome and a clear next step for new users would go far.

Once a system has made a good first impression, it's time to help people get oriented and develop a solid understanding.

Orientation: where am I?

According to legend, in the 5th century BC, the poet Simonides of Ceos attended a banquet at the house of a wealthy nobleman in Thessaly (now Greece). When Simonides stepped out for a few minutes to answer a message, the roof of the banquet hall collapsed, crushing everyone inside. Friends of the deceased wanted to give them a respectful burial, but couldn't tell any of the bodies apart. Simonides realized that he could identify each one by picturing who had been sitting in each place at the table. From this event, he inferred that picturing pieces of information in specific physical locations could be used as an aid to recall. This technique is called the method of loci, or the memory palace. It's used today by real-world competitive memorizers—and the fictional cannibal Hannibal Lecter.

Given how long humans ranged the earth before developing writing, the power of spatial memory makes sense. For millennia, survival depended on remembering the location of fresh water, shelter, or a heard of tasty mammoths. Now we have Yelp. And if you spend a lot of time online, like I do, you might find that your memory seems to deteriorate. I think these things are related.

This phenomenon explains why it's more common to speak about *navigating* websites than *requesting information*. A spatial metaphor makes information easier to retain and process.

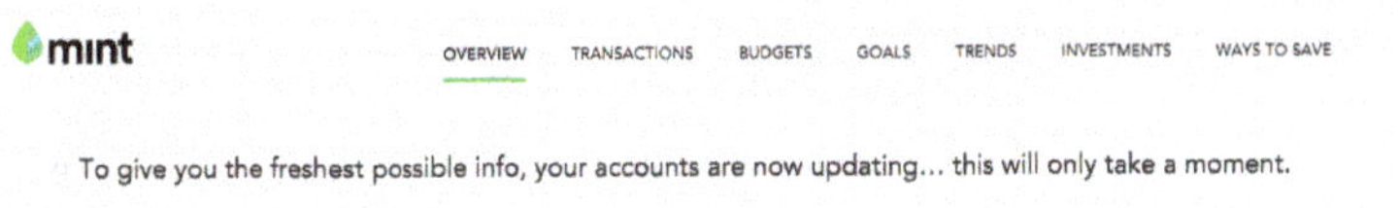
mint
OVERVIEW
TRANSACTIONS
BUDGETS
GOALS
TRENDS
INVESTMENTS
WAYS TO SAVE
To give you the freshest possible info, your accounts are now updating... this will only take a moment.

FIG 3.9: Mint's navigation gets to the point without any attempt to be clever.

Without a spatial reference, it's easy to get lost. If you can use language to create a vivid map in the mind of the user, you can instill confidence and provide ease of use. Even better if your map fits into a pre-existing mental picture and requires no learning.

Navigation as orientation

People glean information from a variety of contextual cues. If you see someone standing behind a bar washing glasses, you can reasonably expect to order a drink from them. If you see someone confidently walking down a city street, you might consider asking them for directions.

According to James Kalbach's *Designing Web Navigation*, structural navigation—the navigation that reflects how an information space is organized—helps users in these ways:

- Expectation setting: "Will I find what I need here?"
- Orientation: "Where am I in this site?"
- Topic switching: "I want to start over."
- Reminding: "My session got interrupted. What was I doing?"
- Boundaries: "What is the scope of this site?"

In a website, navigation labels present the major categories of functionality. They should be as straightforward as possible, as in Mint's labels: phrases like "Budgets," "Goals," and "Ways to Save" are clear and direct (**FIG 3.9**).

Navigation is not the place to try out novel concepts. A cautionary example comes from Hotwired, the first commercial web magazine, launched in 1994 (**FIG 3.10**). I was alive then, and I remember the site, but I don't remember what any of those labels referred to ("Renaissance 2.0," anyone?). The editors

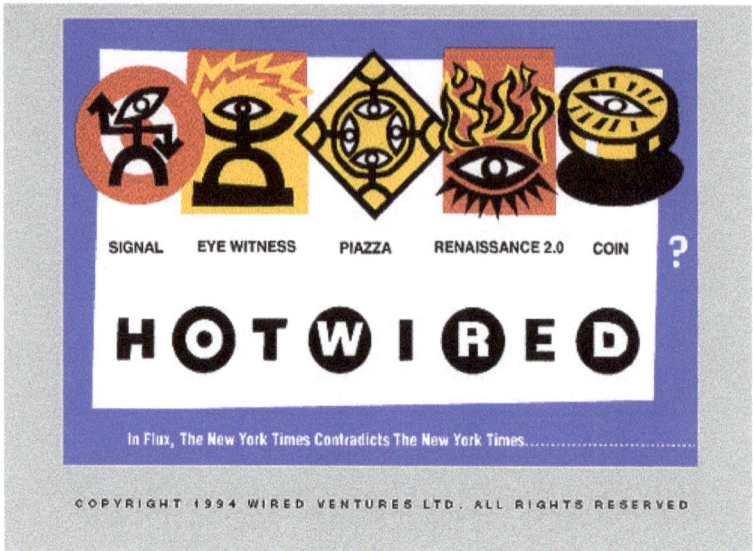

FIG 3.10: Never forget. I'm putting this one in here for the children.

quickly changed them, but the trend of so-called "mystery meat" navigation continued throughout the late 90s. Imagine a highway sign saying, "Crude & Nom-Noms" instead of "Gas & Food."

According to researcher Peter Pirolli, author of *Information Foraging Theory*, people interacting with information systems are doing a continuous cost-benefit analysis, just like animals on the hunt. Everyone wants to know as quickly as possible whether they're on a path to success. And that's exactly what successful navigation does. Surprises and mysteries aren't fun, unless you're writing a thriller!

Context and cues

According to Professor Gabriel Radvansky, a psychology professor at the University of Notre Dame, "entering or exiting through a doorway serves as an 'event boundary' in the mind, which separates episodes of activity and files them away." The same thing happens with information spaces.

As household objects become "smarter," and new modes of interaction emerge and visual cues diverge or disappear, it's critical to rethink how to orient every customer in the information space. Encouraging discovery and habitual use are challenging enough with the standard menu and button GUI, but there is more of an established practice of providing suggestions and recommendations. With voice and messaging interfaces that are intended to feel more conversational, there are missed opportunities to offer a helpful aside.

You need to start with a set of concepts that are familiar and meaningful. People switch between multiple systems, so it's much better to ask how we can orient a service around the user's intention, rather than ask users to reorient themselves as they move from their smartphone to their smart fridge. This is a hard problem. Amazon hasn't solved it, even though people who buy the Echo speaker are likely to be familiar with the Amazon.com ecosystem.

According to a 2016 study by Experian, over 80% of people surveyed reported using the Echo to play a song or set a timer, and there was a sharp drop-off in use after that. Only 32% of users reported using the Echo to buy something on Amazon Prime (http://bkaprt.com/cd/03-01/). There's nothing about interacting with Alexa that cues a user to the full range of possibilities or links one task to the next. Without prompts to interact, customers must be highly motivated to seek out new topics and activities. The Amazon.com recommendation engine is core to the company's success, so it's surprising that Alexa has no suggestions to offer. This may change with time and usage data.

Fundamentally, this is a pitfall of retaining a device-centered perspective on design. Rather than asking, "How do we get our customers to make more purchases using the Amazon Echo?" the better question is, "How can we make the presence of an Amazon Echo in the home provide more value to both Amazon and the customer?" and, "Where do we need to introduce cues and prompts to do that?"

If the designer is clear on the value that the system is providing *and* the context of use, it's possible to identify the best times in the interaction flow to provide additional orientation cues, but possibly in a different mode. As Laura Klein writes in *Design For Voice Interfaces*, "When you think about it, some pieces of information are really easy to say, but they are hard to type, and vice versa. The same goes for output." Maybe when a talking speaker wants to show you the way, the smartphone screen in your hand is the sign.

Choices are work

Navigation represents choices, choices are decisions, and decisions are work. Hick's Law states roughly that the more choices a person faces, the more time it will take to make a choice. This principle is named after the British psychologist William Edmund Hick who created a formula that models reaction time. Offering a quick interaction is one of the most important principles of a conversational design. As with Grice's principles of Quantity and Manner, the most humane thing to do is the work of determining the smallest number of the most appropriate choices to help the customer reach their goal.

Of course, we don't want to sacrifice accuracy. So, the key is to offer the right choices at the right time. Google Search offers the ability to narrow results after the search is conducted, rather than making the user choose up front (**FIG 3.11**).

The order in which a system offers choices is as critical as the number of choices it offers. You need to offer the right number of the right choices at each step to help users make easy decisions and avoid mistakes. The ideal is something like the opposite of an automated phone tree in which the customer proceeds through a series of multiple-choice audio menus that seem unpredictable, banal and interminable. Offer the most consequential or meaningful choices first. For example, booking travel starts with flight, hotel, or car, before getting into destination or dates. Ordering food often begins with the style of cuisine. Solve for the smallest possible number of choices that will satisfy customer intent.

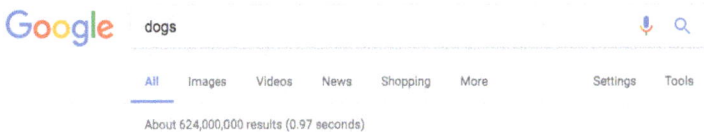

FIG 3.11: Google offers choices on the results page, rather than requiring the user to choose the type of result beforehand. This is much easier to use.

The wrong lesson to take from Google Search is that the open search field represents an ideal way to navigate. An open input—or an open question—represents a nearly infinite set of choices. This is only appropriate when the system offers a large set of potential responses relative to the set of possible requests. Menus may not seem conversational, but they're faster and friendlier than the potentially endless alternative of bad guesses.

Orienting customers in a conversational way means impressing each new mind with the full range of possibilities and then eliminating nearly all of them. This might seem contradictory. In truth, no one with a goal in mind wants to know the full range of possibilities. That's paralyzing. They just want to know if they can accomplish the thing they have in mind *at that moment*. Offering the right choices at the right time is crux.

Seamless, the food delivery service, begins by asking where you live. In this context, it doesn't matter which restaurant is best overall if it doesn't deliver to you. Once the system knows where you live, it asks, "What would you like?" And the arguments with your housemates can begin.

Like a helpful serviceperson, a system that is capable of remembering past behavior can offer appropriate and timely choices based on that information. For returning customers, Seamless offers the option of reordering previous orders (**FIG 3.12**). This suggestion doesn't block finding or choosing a new restaurant, but it's a tremendous convenience for those too tired after a long workday to decide what to have for dinner.

Intelligent digital systems may start making more sophisticated decisions on behalf of their customers, but predicting behavior and preferences is still a tricky business. No one wants

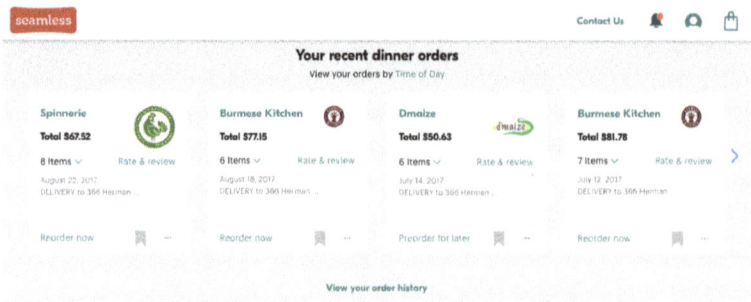

FIG 3.12: By remembering past behavior, Seamless makes choosing dinner easier for repeat customers.

to believe they're *that* predictable. Systems will need to provide at least the illusion that the human is still in control.

Action: what can I do?

I live on Earth at present, and I don't know what I am. I know that I am not a category. I am not a thing—a noun. I seem to be a verb, an evolutionary process—an integral function of the universe.
—BUCKMINSTER FULLER, I Seem to Be A Verb

Reserve a table. Place your order. Tweet. Comment. Pay rent. Kill time. The most fundamental design decision is what your product or service allows your customers to do. Trivial or momentous, the verbs are the point of the conversation, a part (or the whole) of the goal.

Verbs give your system the opportunity to augment the animal joy of physical motion with playfulness, or the exhilarating sense of command (made all the more exhilarating because you no longer need to get off the sofa to order dinner). An action could be anything you can imagine and make happen. This is the heart of interaction design.

With a trivial physical act—a single click, tap, swipe, or utterance—a user can issue an order, triggering significant events across time and space. The actions a system offers allow users to change something in the world, or in their own mental state or that of another.

The hardest work an interface does is translating human actions into machine actions and vice versa. Humans act based on habit, hope, feelings, fears, and sometimes goals, using a mental model patched together from a lifetime of associations. Machines act based on rules and depend upon precision. To ensure a system delivers on the expectations it sets, we often need to break up a single customer action into its component parts and build it back up again piece by piece.

Actions must be goal-oriented, moving the customer closer to their objective. The actions must be context-aware, reflecting the customer's state (in a hurry, using a mobile device) as well as the state of the system (signed-in). Each discrete action must provide feedback. The user needs to know whether their action succeeded or failed and what to do next. The implications of the action must be clear and apparent. As mentioned earlier, truth in interaction is matching expectations. This is the path to trust and credibility.

As in conversation, the sequence of actions should feel cooperative. An unexpected order creates unnecessary work for the user. It's every designer's duty to do the hard work to make the customer's interaction easy.

The better the system matches the customer's mindset and guides action, the more mistake-proof it will be. Of course, things will go wrong. As we know from conversation among humans, it's impossible to prevent all misunderstandings and it's easier to recover in an environment of trust. All actions must be reversible or provide very clear warning when they represent commitments. Any interface for action that omits or misrepresents the full consequences to the user is failing to be truthful.

And since correcting course and getting back on track also requires a set of actions, the same principles apply to helping a user get out of a fix. The goal is to help users recover quickly and clearly while retaining confidence that the system is still on their side.

The context of action

To support the success of the user's action, the system needs to implicitly or explicitly communicate the following in an efficient and context-aware manner:

- **Prerequisites to action.** What does the user need to do before the action is possible? (The user should only be presented with actions they can actually take.)
- **Encouragement to action.** How does the system articulate the benefits of taking the action?
- **Instructions for action.** Is it as simple as clicking Submit, or are we going down a complex path?
- **Consequences of action.** Set expectations of what will happen once the action is taken.
- **Level of commitment.** Is it possible to undo this action?

As an omnichannel platform for separating people from their money, Amazon offers good examples of ways to support action through clear, conversational language. The interaction for watching a sitcom that requires a premium HBO subscription is a study in doing things right (**FIG 3.13**). Amazon offers precisely the right amount of information about the benefits and consequences of completing the action.

The danger of ambiguity

Actions have consequences. It's up to the phrasing of the action to communicate these clearly—the last thing you want to do is make your customers guess at what will happen. Ambiguity decreases trust and increases cognitive overhead—that means creating more anxiety and more work for users.

A trivial example of a European ATM demonstrates how a polite-sounding word—"especially"—can introduce uncertainty into the interaction (**FIG 3.14**). It's trivial because I trust the machine to dispense the correct amount. It's unsettling because why offer a soft preference like "especially"? By pulling the customer out of their flow, this sort of detail introduces anxiety

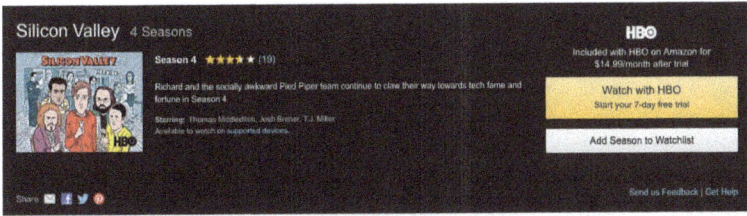

FIG 3.13: All the information necessary to take action is present. The button itself tells the customer that it's possible to watch right away (with the consequence of signing up for a $14.99 subscription).

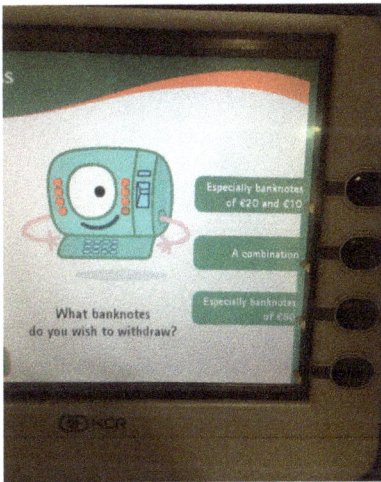

FIG 3.14: This European ATM offers options, but introduces unnecessary ambiguity into the interaction. The word "especially" is either unnecessary or misleading.

and steals time. Even if this is the result of a mistranslation, it's a choice with implications.

Even in a system with nonverbal interface cues or a graphical interface, the choice of action verbs and phrases *is* the design choice (**FIG 3.15**). Even if you're representing an action with an icon, you must be clear about the command that the icon represents. The phrase defines the action. And no amount of styling can overcome misleading wording. Each action has nuanced implications, as well as conventional usage.

FIG 3.15: It matters much more what the button says than how it looks. The phrase defines the action. The style serves only to support this understanding.

The power of choice

As with choices of paths through the information, systems offer a choice of actions to take. Choices not only have a cognitive dimension, but an ethical one too. This is the moment when ease of use, business goals, and honoring the customer's true goals hang in the balance. Any instance of choice may be minor, but they add up. A system working in true interactive cooperation with a user will not lead them down a bad path.

Former Google Design Ethicist Tristan Harris points out that offering a limited set of choices can create the illusion of greater freedom and instead manipulate and hijack the mind of the customer. The given choices may distract from true needs. As he says in "How Technology Hijacks People's Minds":

Erika Hall added a new photo.
September 14 at 11:11pm · Instagram ·

Hopeful Derp

Like   Comment   Share

FIG 3.16: This is an easy post to Like, but what about topics that don't reduce to bland approval or emoji outrage? Action labels define the range of human agency within any particular system. The more people rely on a given system, the more consequential these choices become.

When people are given a menu of choices, they rarely ask:
- *"what's not on the menu?"*
- *"why am I being given these options and not others?"*
- *"do I know the menu provider's goals?"*
- *"is this menu empowering for my original need, or are the choices actually a distraction?"*

Consider Facebook. Facebook mediates billions of relationships around the world: between friends, family, romantic partners, businesses and their customers. No matter the type of relationship or the type of message, Facebook offers a single primary reaction—"Like" (Love, Haha, Wow, Sad, and Angry are hidden, but appear on hover to offer an extended range of one-dimensional reactions).

Some posts are simply likeable, but what about the messages and connections that deserve more than a snap reaction (**FIG 3.16**)? As Tristan Harris posits, these options diminish and obscure the full range of human emotion. There's social pressure to Like, and social acceptance in Liking. These options are so convenient, it's easy to make the argument that Facebook has hijacked how people think about and negotiate relationships.

This consideration adds an additional dimension to the maxim of Truthfulness. It's up to your goals and your conscience to what extent you recognize a duty to minimize distraction from the customer's higher-order life goals. The ultimate challenge for designers is to create a system in balance, one that's transparent about the business goals it represents, while encouraging the user to take those actions that provide value to both the business and the customer.

The feedback loop

Every complex system starts out as a simple system that works—and the simplest coherent system is a feedback loop.
—AMY JO KIM, GAME DESIGNER

Habits are cognitive shortcuts that free our overtaxed brains to think about other things even when we're doing complex tasks, like driving a car. At first, driving a car requires a lot of intense attention. Over time, drivers can steer, accelerate, brake, and check the mirrors without thinking. And—for better or worse—a habitual commuter can get home almost without noticing.

Anyone designing anything for humans needs to consider the point at which a behavior becomes a routine. The traditional dogma of user-centered design is that designers create systems to help people achieve goals by accomplishing tasks. A basic standard for an interactive system is usability—how easy something is to learn and use. Nothing that requires conscious thought will ever be as usable as something that's used out of habit. Habits are hard to compete with, and habits are often entirely unhinged from goals.

A habit is a thing deep in the brain, a feedback loop that forms through repetition. In *The Power of Habit: Why We Do What We Do in Life and Business*, Charles Duhigg popularized the work of MIT researchers to understand the neurological basis of habit. There's a cue—the trigger that sends someone down a habit path—a routine, and a reward. Over time, and through repetition, our brain comes to associate cues with rewards, reinforcing habits.

FIG 3.17: When a habit loop forms, automatic behaviors replace conscious, goal-directed activity. A system can reinforce a habit by providing additional feedback that contributes to a sense of skill building towards mastery, bringing customer goals into alignment with system goals.

Game designer Amy Jo Kim expresses the idea of the habit loop as a type of learning loop, and encourages product designers to consider it in their work. In a learning loop, feedback from the system helps the user get better at interacting with the system over time. This growing mastery is its own powerful reward (**FIG 3.17**).

For example, the core action on Facebook is posting updates. The system gives you feedback in terms of Likes from your friends. Getting more reactions from more people feels like getting better at Facebook, and drives further interaction, whether or not interacting through Facebook has anything to do with any higher-order goal.

When designers talk about "delight," they often mean some unexpected, pleasurable sensation generated from interacting with a product or service. The aspiration is that this delight will be part of the habit-reinforcing reward. It's easy to mess this up by underestimating the amount of cognitive effort an unfamiliar system requires, and by overestimating how pleasurable the reward is. The result can be an unnecessarily novel interface that's hard to learn and leaves the user frustrated rather than delighted. Or an interaction that's only delightful the first time, then increasingly grating through repetition—like your barista telling a knock-knock joke.

Guidance: the system wants you to succeed

In an ideal world, every exchange would be go as smoothly as the perfect dinner party and every offered action would be self-explanatory—but the world is more complicated than that. Everyone has different skills and expertise, and many of the systems we're designing truly do allow their users to accomplish new things in new ways.

Even a well-designed object or system may need to explain itself to help customers succeed every step of the way. Avoiding verbosity may seem more refined, but the confusion is likely to elicit some light profanity in response. A friend of mine photographed this water dispenser at his office. The dispenser required a little aftermarket guidance to facilitate success (FIG 3.18). While this object was designed to exist in mute simplicity, some kind person gave it the power of language to explain itself.

Opportunities to provide guidance include:

- **Sales.** Helping the customer understand whether a product or service will solve their problem or help them meet their goal in advance of any commitment.
- **Instructions.** Any additional information provided in context of the interaction, like stage directions. This general category includes hints and onboarding.
- **Contingency messages.** Letting the customer know something outside the expected flow has happened and getting them back on track. Includes errors.
- **Notifications.** Interruptions that occur outside the context of interaction.
- **Documentation.** Reference materials provided for study outside of interacting.

FIG 3.18: Workplace kitchens are a fruitful location for conducting interface design field studies.

In person-to-person interactions, we talk about offering a high level of service, anticipating needs and providing information at just the right time. Often, the best service is nearly invisible, like a server who manages to keep your drink topped-up without you noticing, or a bank teller who offers a pen along with the paperwork you need to fill out. In interaction design, offering just what's needed for the task at hand is called *progressive disclosure*. The Quartz app does a nice job of this (FIG 3.19), offering a summary of the story that links to the complete article, and an emoji blurb that provides the option to see more detail.

Affordances and clearly labeled actions are not always enough, and that's okay. Often interfaces end up worse off because designers think that a button label or an icon should do all the work. Combining an unambiguous action along with some additional guidance is the best way to support both habitual customers, and those new to the system or infrequent users.

In designing a high level of service-oriented guidance into interactions, there's one thing to keep in mind above all: computers are good at storing and recalling information and people are not. As I've mentioned earlier, computers should do all the

FIG 3.19: Quartz uses
progressive disclosure to let
the viewer dig into the story
as much as they want.

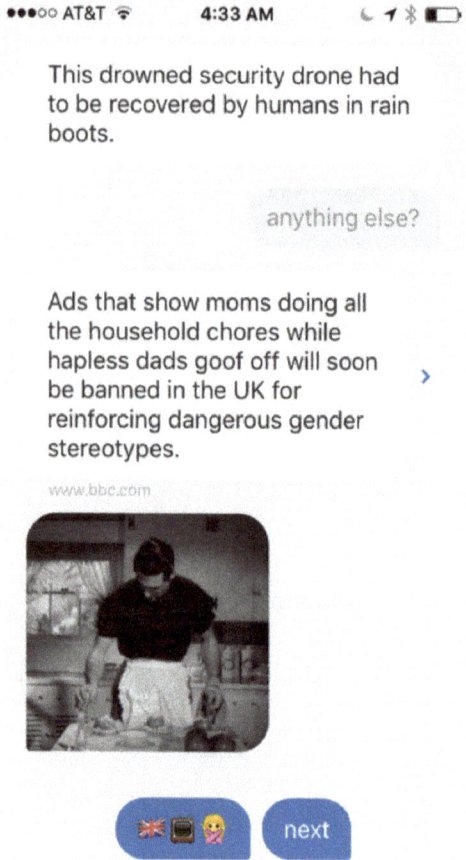

information storage and retrieval in the relationship. If at any point a computer-based system relies on human memory to function, that system has failed its human. In this regard, password-based authentication is the web's most epic failure. This makes password recovery the most "popular" feature of any secure system. Bank of America makes their customers work hard to recover passwords (FIG 3.20). I suspect the designer didn't read these instructions out loud.

Username/Password Help

I forgot my password

For your security, we do not email or present you with your current password, but it's easy to create a new password online. Enter the required information and we'll display your username and ask you to enter a new password.

Get Password Help

I forgot my username

We can remind you what your username is. Enter your Social Security number and your password, and we will display it for you.

Get Username Help

FIG 3.20: Bank of America shows how adding more explanatory text can be less helpful. The user knows whether they forgot their password or their username. The detailed instructions belong after the choice, not before, violating conversational maxims. This is too much information at the wrong time.

Your Memories on Facebook

Erika, we care about you and the memories you share here. We thought you'd like to look back on this post from 2 years ago.

FIG 3.21: Facebook walks a fine line between being service-oriented and presumptuous, and sometimes crosses it, revealing that they're watching your every move.

While you should seek every opportunity to use stored information to benefit customers, also do your best not to creep people out. This is something Facebook has trouble with (**FIG 3.21**). Facebook tracks interpersonal interactions across billions of relationships, along with all the associated data users add to the system. By referring to themselves as a "we" that has concerns about "you," Facebook destroys the comforting illusion that it's just a neutral utility.

Notifications

It used to be that you had to interact with a system through its interface to exchange information with it. Now, thanks to push notifications, a system can get your attention even if you're not currently using it.

Notifications do serve a purpose. In theory, they allow a system to continue to serve a customer without requiring continuous attention. In practice, they're easy to abuse. The desire for engagement can cloud judgment about what sorts of interruptions actually benefit the customer. The disruption can snowball, especially if you have relationships with multiple systems, or the same system across multiple devices. The proliferation of notifications can lead to constant interruption, which is the opposite of feeling served and empowered by technology.

In general, notifications should require affirmative consent from the customer. Therefore, you need to establish trust and offer value before prompting your customer to receive notifications. It's particularly important to approach notifications from a place of politeness, and have a holistic strategy across all messaging types and channels that the system uses to convey information.

The characteristics of helpful notifications:

- **They're well-timed.** In the old days, the worst notification was a phone call during dinner. Notifications should arrive at the right time for the customer to respond. Unless the notification is for an early morning alarm, 3 am is unlikely to be that time.
- **They're concise and clear.** This goes for all communication, especially for alerts requiring action.
- **They're personalized and relevant to the customer.** Unless the notification is truly an emergency, use other forms of communication for general messages.
- **They deliver value and enable action.** Notifications should only be used to alert the customer to something they need to take action on. Creating a sense of urgency when there's no possible action just creates anxiety. No one needs more anxiety.

- **They generate interest and reward trust.** Every notification is a potential enticement to shut off all notifications. Don't be that.

Google's Material Design Guidelines is written for Android App designers, but its Notification guidelines are broadly applicable (http://bkaprt.com/cd/03-02/). They say that notifications are appropriate for:

- communication from another person through your system, assuming it's a person your customer is likely to want to hear from;
- action that helps your customer meet a goal or have a better quality of life, such as getting to a meeting on time, or avoiding rushing to make a flight that's been delayed; and
- a system state change that suggests or requires action.

In summary, notifications are appropriate only when it's likely the customer will welcome the interruption and have the ability to take action in response, such as acknowledging a calendar event or shutting off an alarm.

Notifications are not appropriate for:

- **Advertising.** Unless the customer has opted-in specifically to promotional messages, avoid interrupting them with a commercial.
- **Messages with no customer value.** Be very clear about the value of the service you're providing, or don't create a notification.
- **Situations in which there is no action.** If there's nothing for the customer to act on, there's no reason to interrupt them.

In 2014, Skype demonstrated a high degree of context awareness by introducing the concept of an "active endpoint." If a Skype user is signed into multiple devices, say a laptop, tablet, and a smartphone, but they're only sending chat messages from the phone, Skype chat notifications will only go to the phone. All the other devices will remain silent, at least with respect to Skype.

Onboarding

The onboarding process is how you help a new (or long-lost) customer feel at ease, in control, and productive. The term comes from the common human resources practice of new-hire orientation. Just dropping someone cold into a new organization and leaving them to figure things out on their own is unkind. No one likes to feel ignorant or inept. It's in the interest of an organization to help new staff members feel competent and supported. The same goes for new customers.

The amount and type of onboarding guidance your system needs depend on three things:

1. How different the system is from others your customer may have used already
2. The conceptual complexity of the system
3. The amount of effort the customer needs to put in before getting something useful out

A well-designed system should draw on the existing mental models of potential customers and require little effort before payoff. However, if your system presents unfamiliar concepts—a new way to accomplish something or a way to accomplish something complex—you may need to design an experience specifically for new users. This onboarding experience should be integrated into the overall experience, and be as lightweight as possible—the opposite of a long, up-front tutorial. For example, Duolingo does a good job of helping users self-sort based on their expertise and desire to just get started (FIG 3.22).

To design effective onboarding, you need to know which customer actions will deliver the greatest value. And if you don't know what constitutes value, before designing the system, you'd better do some user research.

In general, the best onboarding is the least intrusive. Don't focus on creating a delightful process that's an experience in and of itself. Focus on getting the user to the value in a way that supports your business goals. Identify barriers to achieving value and what the customer needs to overcome them, whether it's

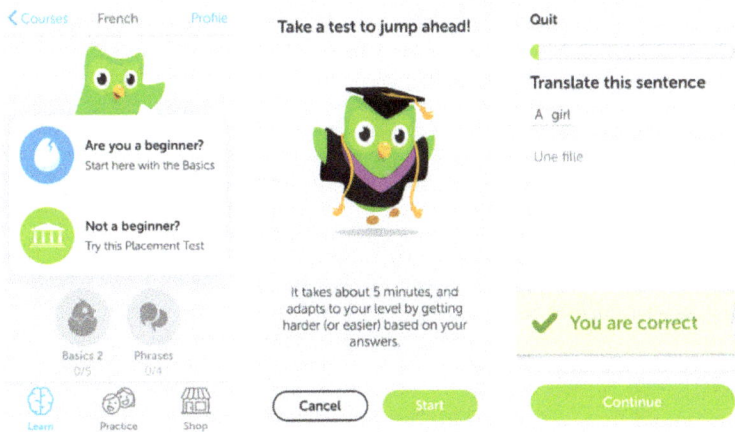

FIG 3.22: Duolingo offers multiple paths to get started and explains the potential benefits of each. Also, the time commitment is explicit.

simple inline information, or hand-holding and encouragement throughout the process.

ACCIDENTS HAPPEN

The designer shouldn't think of a simple dichotomy between errors and correct behavior: rather, the entire interaction should be treated as a cooperative endeavor between person and machine, one in which misconceptions can arise on either side.
—DONALD NORMAN, The Design of Everyday Things

We designers demand creativity of ourselves, but don't anticipate it in the people we design for. That's a failure of imagination. Expecting that people will behave "correctly" is the path to fragile interactions.

Humans are not precision systems. Our memories lapse, our fingers slip, and we do things in the wrong order. Screwing up is our birthright. Science fiction has enshrined in us a sense of superiority—some special human quality is often the proffered explanation for escape from an extraplanetary scrape—but it's really our mistakes that define us.

How a machine responds to human error is the best test I can think of for its ability to mimic human intelligence—or at least to seem more human than cold, calculating machine. Precise calculations and dazzling displays of information retrieval are easy for machines, but supportive redirection? Not so much. Forgiveness may be divine, but it's certainly not digital.

Humans are improvisational; machines run according to rules. From the perspective of a computer, what is human error? It's an unanticipated input that leads to dead ends and unnecessary frustration (for the human, not the machine). Too many system responses, even in sophisticated interfaces, are simply variations on "Does not compute."

Errors in human social interactions provide opportunities for the unexpected, both tragic and delightful. When humans make mistakes with each other, the most humane thing to do is to gently correct, redirect, and try to prevent the error from happening again in the future.

What if the problem is the system's fault? Sometimes the human does everything right, but the machine makes an error and the system breaks. With some clever branding, this can be made to be charming, like the early Twitter fail whale (**FIG 3.23**) which turned server outage into a cultural icon. What isn't charming is being too glib if the system failure causes a serious setback, especially anything involving the loss of personal data. Imagine your doctor setting your medical records on fire and then just shrugging with an "Oops!" A sincere apology and a quick path to setting things right would be more appropriate.

Matt Jones, interaction designer and author, coined the phrase "be as smart as a puppy," (also affectionately referred to as *BASAAP*) to say designers should be "making smart things that don't try to be too smart and fail, and indeed, by design, make endearing failures in their attempts to learn and improve. Like puppies" (http://bkaprt.com/cd/03-03/). This is a useful notion to consider in terms of setting expectations for how much intelligence you can expect from a system that's learning from its user over time. However, people don't typically depend upon

FIG 3.23: The illustration of a whale being lifted by birds was originally called Lifting a Dream.

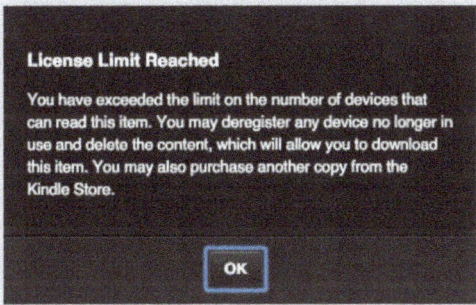

FIG 3.24: This is a hostage situation. The user is offered no alternative but to acquiesce.

puppies to help them accomplish important goals. Failure is not endearing to someone trying to book a flight or make a money transfer. Or, trying to read a book.

The Amazon Kindle surprised me by letting me know I couldn't read a book I had purchased sometime previously (**FIG 3.24**). This message presumes the customer knows how to deregister a device. There's no path to action, just an acknowledgement of the roadblock.

Do the poka-yoke

Poka-yoke is the Japanese term for mistake-proofing in manufacturing. Poka-yoke designs are most common in machines and devices that are dangerous if used incorrectly. For example, no microwave will start unless the door is shut.

The simplest poka-yoke in digital interaction is only allowing user input within boundaries. For example, by offering a menu of options rather than an open text input, you can mistake-proof an online form. It's an excellent design principle to run through every possible scenario in which a customer might use a system as designed and yet end up harming themselves, like misspeaking or misspelling the names of their medications.

From a business relationship perspective, email is treacherous. Prevent your customer from looking like a careless idiot, and you earn their loyalty forever. Gmail offers a fantastic poka-yoke feature (FIG 3.25). If you try to send a message that contains a phrase such as "attached to this message" or "I've included" without attaching a file, a dialog box appears to alert you to your error. Now if only they could do something about the toxic waste spill that is reply-all!

Interaction design begins in the mind of the customer, which means error prevention starts in the first moments of creating the awareness that any given system exists. Make sure you start with the right concept, one that matches the users' mental model. You cannot control user expectations and associations, so you must understand them and look beyond your product to see what creates them. Assume everyone is on autopilot all the time, and that they're drawing unconscious inferences from the barest of cues. Asking for open-ended input will lead to the assumption that any answer is acceptable. A voice that sounds too much like a person will set expectations too high.

Our human brains are great at many things, but we often have as much trouble remembering as predicting. Never require users to retain information in their memories. Always ask, "Are you sure?" before dangerous or consequential actions. And warn people, gently, when they're getting too close to the rails. Like elephants, machines have excellent memories. Tax the robots, not the apes.

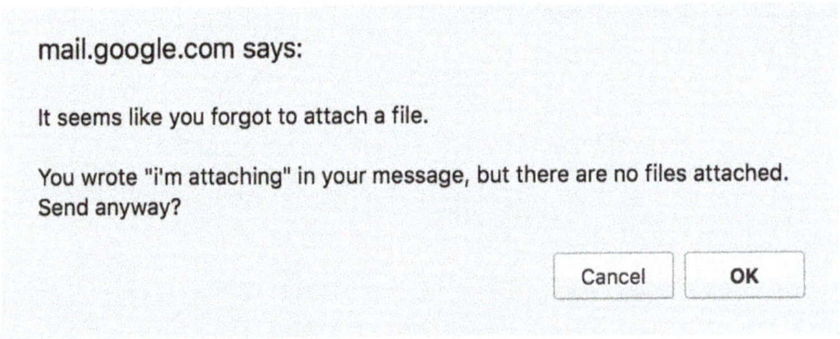

> **mail.google.com says:**
>
> It seems like you forgot to attach a file.
>
> You wrote "i'm attaching" in your message, but there are no files attached. Send anyway?
>
> Cancel   **OK**

FIG 3.25: Gmail has saved me from myself countless times.

CONTEXT MAKES THE CONVERSATION

As interconnected digital systems endeavor to offer more "natural" ways of interacting through voice and text, the limitations of these systems, combined with the context of their use, can make interacting with them somewhat of a minefield. This leads to the sort of frustration intuitive interfaces are supposed to prevent. Being able to interact with a computer in the same way you text a friend or talk on the phone sets high expectations. Human and machine may be conversing, but they are not cooperating. For example, the H&M shopping assistant on the Kik messaging service creates a much more awkward interaction than simply browsing a website (**FIG 3.26**).

Natural language processing requires the computing power to analyze and interpret human speech or text in real time. There's no room for ambiguity.

Before rushing to chat, consider whether it really will make life easier. Some of the drawbacks include:

- **Lack of context awareness.** The systems can't pick up on contextual cues that might be available to a human, and probing for information would cross over into uncanny valley. Siri or Alexa could ask, "What are you doing right now?" to help train it to be more context aware, but that would be creepy.

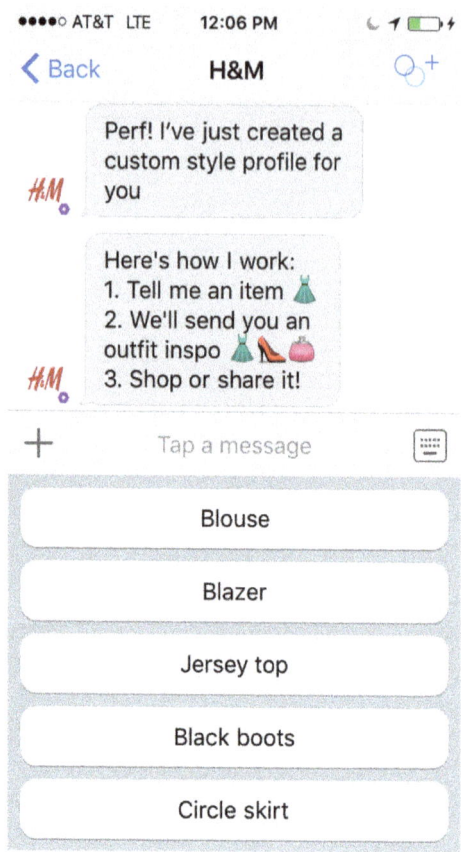

●●●●○ AT&T LTE   12:06 PM   ⌇ ⁄ ▭ ⁺

< Back   **H&M**   ⦵⁺

> Perf! I've just created a custom style profile for you

H&M

> Here's how I work:
> 1. Tell me an item 👗
> 2. We'll send you an outfit inspo 👗👠👜
> 3. Shop or share it!

H&M

+   Tap a message   ⌨

Blouse
Blazer
Jersey top
Black boots
Circle skirt

- **Takes more time.** The web has made self-service fast and satisfying for a wide variety of complex tasks. It's easy to become habituated to speed. Talking through a task can feel agonizingly slow, much more so than clicking a button.
- **Unpredictable.** Even the most intelligent system is vastly more limited than a human. It can be impossible to predict what options are available or what input is acceptable.
- **Not error-tolerant.** Unexpected input can bring interactions to a halt with no path forward.

NOW, LOOK WHO'S TALKING

While it can be helpful to analyze an individual's interaction with a system into discrete moments, for the sake of conceptual clarity in the design process, they're all intertwined. The same conversational principles apply across all moments in an experience, just with different emphases at different times. And all customers will have slightly different needs and preferences; some want to take action before really understanding who they're dealing with, while others are more careful before diving in. Avoid making assumptions about what users want in favor of understanding real-world contexts.

A set of conversational principles that can be used in developing speaking interfaces is a good starting point, but it's the personality that unifies the experience and brings a conversation to life. A personality is a set of stable characteristics, manifested in behavior patterns and situational responses. We tend to consider personality traits as constant and meaningful (regardless of fluctuations in mood). If your friend tells you about meeting a new person, you may ask, "What are they like?" and expect an answer in adjectives—energetic, withdrawn, friendly, creepy, talkative, or laid-back. Words that describe their personality help you understand what it's like to interact with them.

Conversational principles are nothing without a unifying personality—the animating spirit. A distinctive, appropriate, consistent personality brings the principle of conversational design to life. The more sophisticated the interaction you can have with a system, the more personality plays a part. After all, a golden retriever has more personality than a goldfish. Let's look at how language and behavior can demonstrate the personality of a product or service.

4

THE POWER OF PERSONALITY

CHARLIE KAUFMAN'S *Anomalisa* employs creepily realistic stop-motion puppetry to tell the story of Michael Stone, a customer service expert attending a conference in Cincinnati. With one exception, every person Michael encounters—hotel staff, conference attendees, airplane passengers—makes monotonous chitchat in an identical voice. This is conversation bereft of personality. A world of undifferentiated interaction closes in.

In life, every one of us manages to have a unique personality without even thinking about it. But imbuing a service with a personality requires a lot of thought. Creating and sustaining an appropriate and effective human voice within an interface requires effort and intention. It's all too easy to neglect sociability and let the machine do the talking, or overwork your words into chipper clichés.

Because of this, the public voices of so many organizations sound the same, even when they're talking about very different things. It might make sense to avoid being too idiosyncratic if you're trying to reach the broadest possible audience. But if you're truly offering something of unique value to the world, the way you talk about it should have a distinct quality, right?

Otherwise your customers will find themselves trying to navigate a mid-level netherworld where everything sounds like the same pair of beige slacks. And no one wants that.

YOU'VE GOT PERSONALITY

A personality is the consistent set of human characteristics embodied by your product, service, or organization. While a brand is the sum of all the associations in the mind of your customer, the personality is how the system is designed to sound and behave. Voice and personality are often used as synonyms, but even a wordless interactive system may have a personality, as long as it displays or elicits emotions that are sufficiently human. (That a text can "say" something important, and one author may "sound" like another in writing, shows how easily we've been mentally switching modes all along. Now that more of our devices are speaking aloud, it's probably better to reserve "voice" in practice for the audible aspects of an interface.)

In addition to a brand and personality, some companies also create a character. A character is a named entity apart from your organization or brand name that takes the personification a bit further and represents some or all of the brand attributes or characteristics. Organizations sometimes use characters as mascots to add playfulness and increase recognition. For example, MailChimp has a cartoon chimp named Freddie who functions as a simian emoticon. Freddie smiles and winks, but never speaks.

Anthropomorphizing is a thing people do. We see faces in electrical outlets and talk to our plants. So, if you don't craft a personality intentionally, one will be assigned by your customers, in their minds. And it won't be as good and the one you create and control.

One of the most effective and delightful personalities belongs to Slack. I'll let Anna Pickard, the editorial director, explain:

> *There's a sense of recognition, the same simple, straightforward language that helps you in the onboarding process is the one that carries you through every interaction. It is sometimes funny, sometimes serious, sometimes just plain and informative, but throughout, it should feel like nothing more than a person, talking to another person. Human to human.*
>
> *The voice and design are woven tightly together. The bright colours of the logo and playfulness of animations match the tone of the copy in the product and all around it. The tone was set early on – it's very much in the voice of Stewart Butterfield (and the other founders) and my work has been in learning how to scale and expand this without losing the sense that you were talking to a friendly co-worker. (http://bkaprt.com/cd/04-01/)*

The fewer senses your service engages, the more work falls to the voice. Chatbots, the text-based agents that populate messaging systems, are nothing without their personalities. As Ben Brown, cofounder of botmaker Howdy, said:

> *It's nearly the ultimate challenge for digital design, because in most cases, you don't have control of what it looks like at all. How can you boil your entire app experience down into two lines of text? There's nothing else on the screen but that. (http://bkaprt.com/cd/04-02/)*

If your product communicates with a voice UI, not only does personality carry the weight of the product design, you'll need to assign a gender (or offer a selection), as well. This introduces a whole range of considerations and opportunities. Customers may either make an emotional connection, or be repelled based on their associations between a system's capabilities and its gender presentation.

Both Amazon and Google offer voice-activated talking speakers (Amazon Echo and Google Home respectively) for use as home assistants. Amazon named theirs Alexa, while Google refrained from creating a separate named personality. As Google's Scott Huffman put it, they "shied away from the idea of kind of a human persona for search or for the entity that you're interacting with and instead tried to go for, in some sense, 'hey, you're interacting with all of Google.'"

The implications of this choice prompted the designer Johna Paolino to write a Medium post reflecting on how it affected her:

> Amazon is the name of a pioneering e-commerce platform and revolutionary cloud computing company. Echo is its product name, first to market of its kind. Alexa? Alexa is just the name of a female that performs personal tasks for you in your home.
>
> Apple & Siri set a precedent for this. Was there a need to rename the voice component of these products? Why isn't it Echo or Amazon? Why not Apple? By doing this, we've subconsciously constrained the capabilities of a female. With the Echo, we've even gone as far as to confine her to a home.
>
> The voice component of the Google Home however is simply triggered with "Google". Google, a multinational, first-of-its-kind technology company. Suddenly a female's voice represents a lot more. This made me happy. (*http://bkaprt.com/cd/04-03/*)

So, Google's choice to make the personality resemble a human-less individual made an actual human individual much happier. This process clearly requires a nuanced approach.

HOW TO BE YOURSELF

The great enemy of clear language is insincerity. When there is a gap between one's real and one's declared aims, one turns as it were instinctively to long words and exhausted idioms, like a cuttlefish spurting out ink.
—GEORGE ORWELL, Politics and the English Language

Creating a consistent, appropriate personality that expresses itself clearly and elicits the right emotional response in the right people isn't magic—it's mostly science, with a little bit of art. To make an emotional connection with people, you need to understand those people and what they're emotionally connected to. Otherwise you'll create the personality you want to be friends with, and you'll convince yourself it's who they want to be friends with too. We're all very good at rationalizing our choices.

To create interfaces that are meaningful to actual humans who have no relationship to your company, you must listen to them. This means doing user research before starting the design work, and continuously as you create and refine your interface and personality.

This is not optional. You need to hear how real people talk about the specific services you provide, and how they talk about their day and see their problems. Your goal is to hear and understand what your users value and how they talk about it—not so you will mimic them, but so you will be intelligible to them. You must rid yourself of your internal phrasings and find new ones that seem natural to your users. This is the core of the challenge—using language to influence the emotional state of your customer in different contexts and at different points in the interaction. You need to understand their full range of potential emotional states and contexts, and acknowledge and anticipate the negative emotions. Consider the anxious, confused, overwhelmed and skeptical, the bored, the hangry, and the frustrated. Only by working through the potential negative scenarios with other people will you find the right balance of clarity and humanity to help them connect.

Listen for their tasks, too, of course—those things they expect to use the system to do. You need to understand their goals and aspirations and how what you're trying to do fits into their lives. How does interacting with your product or service help someone be the person they want to be? And how do you prove that this is the case?

In addition to just plain eavesdropping, conduct interviews and listen to your users' language. Ask representative target customers about their typical day. Just be quiet, and let them speak. Do this a half-dozen times. Then go back through your notes and pull out all the nouns and verbs. This will tell you what your customers do, and the words they use to describe what they do. As you develop the personality and vocabulary for your interface, including labels and phrases for actions, use these notes as a reference. The goal isn't to sound *like* a customer; your interface isn't a peer. The goal is to be meaningful to your customer and trigger the right set of associations.

Learn to like people

To script interactions between humans and machines, it's helpful to look at how screenwriters script dialogue. My friend, screenwriter and all around good guy Josh A. Cagan, told me that the secret to writing dialogue lies in practicing the difficult discipline of loving your fellow humans:

> *You have to like people. You can't hate people and do your job effectively selling ideas and concepts to people. Of course all of us get sick of other people, but I think people are good at their core. They have ideas that don't jibe with mine and they try to get what they need in bad ways. The secret sauce is that it's awesome to have people psyched about the things you're excited about in different ways for different reasons.*

Adopting this attitude and making it a mantra will make a difference.

I love sandwiches. I've had jobs making sandwiches. I can tell when a sandwich has been made by a sad sandwich maker who has little love for the craft or for the customer. The result is slapped together and unbalanced. The same is true of interactive design. You can tell which products are made by organizations who talk about people, and those who talk about "eyeballs" and "uniques." A cynical approach to people results in impersonal, system-centered design with a slick veneer of marketing.

And as Josh told me, liking your customers and valuing them doesn't mean identifying with them. It means you appreciate that they have a different perspective. You must care about them, or you can't expect them to care about you.

Clarify your values

You need clear values to create a personality with integrity. Values sound like something high-minded and abstract, something that calls on bureaucratic lyricism. But every business has a starter set. Values are implicit in the business model, and denote the exchange of value between the business and the customer. Unexamined values tend to come from an attitude of "we're here to make money doing stuff." This leads to a bland marketing tone that won't stand out. And you can't adopt values that run counter to how you make your money.

Defining and clarifying values is an activity that requires the input of key organizational stakeholders. And it must take place in the context of creating a living interaction between the system (representing the organization) and the customers. Personalities go wrong when the people at the top write up a set of abstract core values or brand guidelines that eventually find their way to a design team where they must be interpreted into a living, interactive experience.

Start with your values as a company and put them in human terms. Unlike many mission statements, these statements should sound like something a real person would say in conversation—that's where you'll find the energy and life. To do this, gather key people in a room with a whiteboard, or over a video conference, or on Slack, and have everyone fill out a mad lib (FIG 4.1).

This can also work as a workshop-style exercise in which small groups discuss and fill in the sentences together, then have a larger group discussion about the answers (FIG 4.2). The important thing is that it's a collaborative process subject to an open discussion, rather than something handed down from on high.

Once you've clarified your values with that exercise, you can start thinking about how your product or service fits into the lives of others. Another mad lib template can work well here (FIG 4.3). Think of the qualities that describe you, and those that you admire and aspire to. Think of the ones that truly differentiate you (FIG 4.4). This can help you recognize that your organization may mean something different to your

We will be successful when _____.
                                    outcome

We care about _____ because _____.
              idea or thing   *reason*

We care about _____ because _____.
              idea or thing   *reason*

We care about _____ because _____.
              idea or thing   *reason*

FIG 4.1: The blanks in these mad libs prompt key people in your organization to frame values in human terms.

We will be successful when **we provide the soundtrack to our subscribers' lives.**

We care about **music** because **it connects us all.**

We care about **how people feel** because **music is emotional.**

We care about **offering all types of music** because **taste is personal and discovery is exciting.**

FIG 4.2: This is what the final result looks like if a music streaming service like Spotify were to fill out a values mad lib. Straightforward, easy to understand, and human.

customer than it does to you. As the maxim goes, you are not the user—forgetting this is the path to a forgettable outcome.

As you and your team design your system, ask yourselves if it sounds and behaves in accordance with these values and qualities.

After doing this exercise, with your team, make a list of all the words you want your audience to use to describe you. Then make another list of all the adjectives you want to avoid. Narrow your list down to three positive adjectives that are unique to you, and three that concern you the most. Use those words to guide all your work.

If we were a person out there in the world serving our customer, our job would be

_____.

the primary role your product plays

And customers would describe us as the most _____,

adjective

_____, and _____ of any in that profession.

adjective   *adjective*

We never want to come off as _____, _____, or

negative adjective   *negative adjective*

_____.

negative adjective

FIG 4.3: Another exercise asks people to think of their organization as a person, identifying adjectives that can serve as potential personality traits.

If we were a person out there in the world serving our customer, our job would be **DJ/music librarian.**

And customers would describe us as the most **savvy, eclectic,** and **perceptive** of any in that profession.

We never want to come off as **snobby, stale,** or **narrow.**

FIG 4.4: If an organization like Spotify were to fill in those blanks, they might personify their service as a savvy-never-snobby music librarian.

If you focus on what you have to say before thinking about how you sound, you'll have an easier time sounding right. Think of it as drawing the right style out of the subject matter rather than attempting to apply authenticity like a veneer. You need to know who you are to accomplish this. Otherwise, it's easy to fall into abstractions. Lead with the functions and flow that reflect what your customers are already concerned about and you can inhabit existing neural pathways rather than carving new ones.

Know your role

Some organizations are concerned about being "too conversational," by which they mean "too informal." But being conversational doesn't imply anything about the seriousness of the conversation or who is having it. Doctors have conversations. Bankers have conversations. Funeral directors have conversations. Most of these conversations happen well within the bounds of professional propriety.

If you're designing a banking system, it should sound like a reliable, helpful banker. A funeral planning system should sound like a compassionate funeral director. And if you're designing a game with a shrewd adversary, it should sound like a malevolent, self-important synthetic intelligence (**FIG 4.5**).

There are plenty of exercises that can help you imagine your product's personality. *What kind of car would your application be? Or is it a tree?* These kinds of exercises can be a fun way to get the conversation started but often feel too abstract to be applicable to system interactions.

My own research has shown that if you try to determine which celebrity your interface should emulate, it always comes up with George Clooney! So, don't even bother—just figure out the actual human analog for the role you play in the lives of your audience. Real estate agent? Maître d'? Bike shop mechanic? You'll want to streamline the language a bit to be appropriate to use online, but this will give you a good starting point.

For instance, a skate shop can adjust their jargon to be more welcoming to people who wander in off the street as well as to hard-core skaters—it would be jarring if a skate shop sounded like a bank. Or the other way around:

Our mortgage refi options are sick!

There are alternatives to the twin poles of technocratic and too cute. The fundamental characteristics of the personality

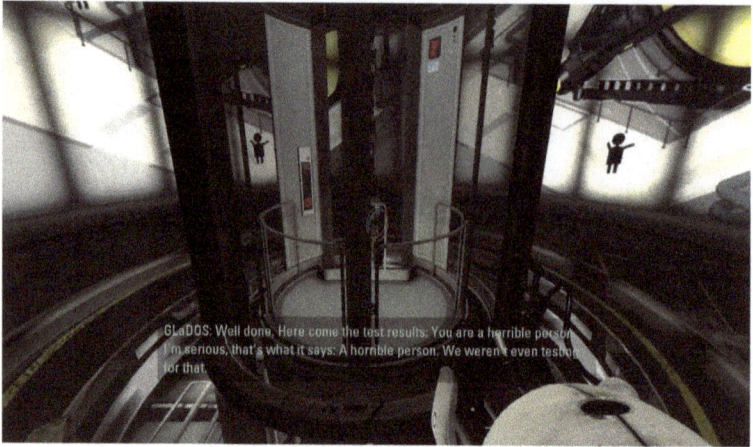

FIGURE 4.5: GLaDOS (Genetic Lifeform and Disk Operating System) is the character of a sinister and witty AI antagonist in the videogame Portal that sounds like synthesized text-to-speech, but it's voiced by a human actor.

should be stable. However, personalities adopt different tones depending on subject matter, and context.

- **Identity.** Is the product the face of the organization? If not, how are they related? When customers interact with the system you're designing, are they interacting with the company, the service, or a named agent? For example, Amazon offers interactions as Amazon, the ecommerce platform, Kindle, the ereader with associated services, and Alexa, the named agent accessible through the Echo speaker. Google Home's Google Assistant takes a different approach, lacking gender and a separate identity. The degree to which the system exhibits its own identity sets expectations for functionality and level of familiarity.

- **Expertise.** How much should the user expect the system to know, and about which topics? A system in the role of a real estate agent should know about mortgages and the housing market. A banker should know about interest rates and savings accounts. A system in the role of a transactional bank teller will express expertise differently than a private banker or financial advisor.

- **Mood and attitude.** In *Hitchhiker's Guide to the Galaxy,* Douglas Adams spoofed the idea of giving computer interfaces human personalities. One of Adams's most famous creations is Marvin the Paranoid Android. Marvin's a severely depressed robot, the product of the Real People Personalities designed by the Sirius Cybernetics Corporation, and a cautionary example of making computers all too human. Most actual interfaces are neutral and slightly positive; the most common mood they demonstrate is some variation of cheerful. Apple's Siri can display a bit of snark in its responses. It's worth defining this explicitly, especially if you want to have a more human personality.
- **Relationship.** Do you see the system as an advisor, a teacher, an assistant, or simply as a tool? The greater clarity you have about the relationship you have with your customer, the better you'll be able to create a cohesive personality that offers the right cues. Again, many systems will include aspects of multiple relationships.

PERSONALITY IN ACTION

Aspects of the personality's role and attitude can change with different parts of the system—just as a customer might encounter different representatives of the same company. At various points, it will be natural for the system to sound more like a supervisor or a helpful customer support agent. This will set different expectations from talking to a sales person (or a malevolent intelligence!).

It's in the mundane details of the interactions where a robust personality can really make a difference.

Tiffany & Co. is a very old and famous jewelry brand. Tiffany sells a very particular style of luxury gift, and have a clear sense of their role in their customer's life. The core customer interaction is selecting and purchasing a gift. On their website, they offer a very Tiffany-style way of allowing a customer to signify to someone their desire for a potential gift (**FIG 4.6**). The inter-

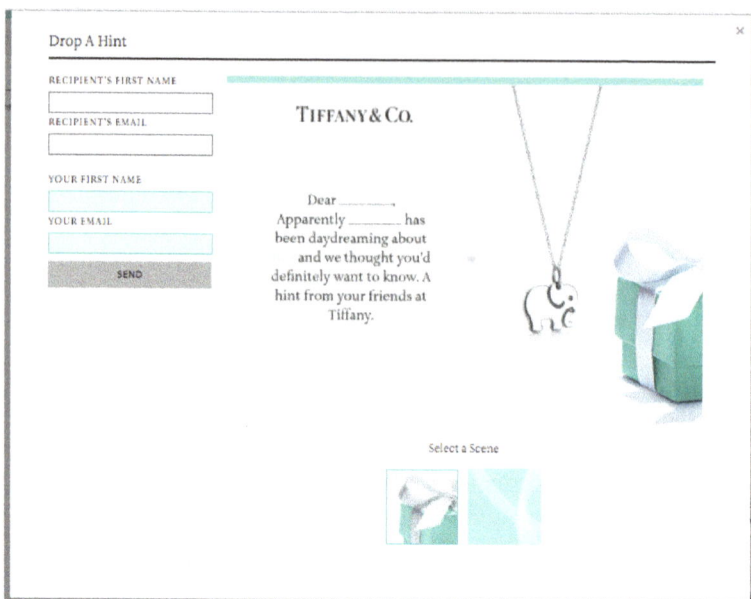

FIGURE 4.6: The role Tiffany plays is to (stylishly) intercede between potential recipient and giver.

action is presented as a hint the store just happened to pick up from someone daydreaming. A generic "email this to a friend" message would come across as crass and, ahem, tarnish the brand. On Amazon.com, a Tiffany-style message would seem fussy and overbearing.

Personality pitfalls

When the personality of your system depends on language, and you have an international audience, it's necessary to adapt that personality to the local language and culture. Simply translating for meaning won't be effective. The marketing and advertising industry figured this out in the 1960s, and coined the term *transcreation*. Transcreation is about creating a message that evokes the same emotions and carries the same implications from one culture to the next.

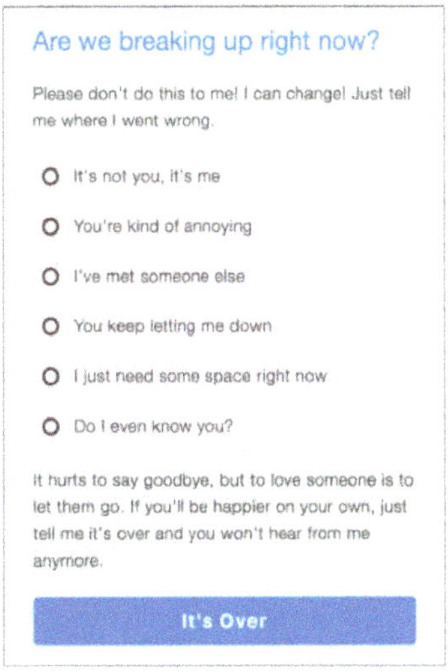

Are we breaking up right now?

Please don't do this to me! I can change! Just tell me where I went wrong.

○ It's not you, it's me

○ You're kind of annoying

○ I've met someone else

○ You keep letting me down

○ I just need some space right now

○ Do I even know you?

It hurts to say goodbye, but to love someone is to let them go. If you'll be happier on your own, just tell me it's over and you won't hear from me anymore.

It's Over

FIG 4.7: Some design decisions are mysterious. This is definitely not appealing to anyone who's ended a difficult relationship.

The Amazon Echo is a personality to watch as it evolves to meet the needs of new markets. Amazon has dubbed Alexa's more idiomatic phrases, "speechcons":

In February, we announced SSML support for speechcons in the US. Speechcons are special words and phrases that Alexa expresses in a more characterful way, making the user experience even more engaging and personal. Today, we are excited to announce that developers in the UK and Germany can now use speechcons to build more creative voice experiences with UK English and German words and phrases. This means you can now build UK English and German Alexa skills that pronounce common words in a more natural manner. You can use regionally specific terms such as "Blimey" and "Bob's your uncle," in the UK and "Da lachen ja die Hühner" and "Donnerwetter" in Germany. (http://bkaprt.com/cd/04-04/)

Alexa also now has the ability to speak in a whisper, ostensibly to sound more human. I have listened to the whispering, and I can report that, as of this writing, it's about as creepy as stop-motion puppets wearing the same lifelike face.

Missteps can also be instructive. At every point, in every role, your interface should come across as supportive and on the side of the customer, even during less positive interactions. Don't do what this financial software did (**FIG 4.7**) and put yourself in the role of a clingy romantic partner. That is not cute, or adult. No one wants that.

A FEW MORE WORDS ABOUT WORDS

To find the joy in fellow humans, and to give it back through systems that are sufficiently humane, it helps to rediscover the joy of language. It also helps to banish the banal habits that begin to rub off from frequent exposure to worst practices.

Our greatest accomplishments as verbal animals include jokes and poetry. Our most shameful sin is soulless corporate jargon.

Good humor

A user interface is like a joke. If you have to explain it, it's not that good.
—MARTIN LEBLANC, founder of Iconfinder

Humor is a critical dimension of human communication and we still don't know exactly why it exists or how it works. There are several theories—emerging from various academic fields, particularly in education and instructional design—that attempt to explain it. Teachers are especially interested in the challenges of holding attention, and imparting information and humor seems to help both.

Humor relieves psychological tension and can help resolve incongruity between an idea and a situation. Over the course of human evolution, I figure enough of our ancestors escaped death by doing something unexpected and delightful, so perhaps we ended up with genes for that. The bores must have relied on camouflage!

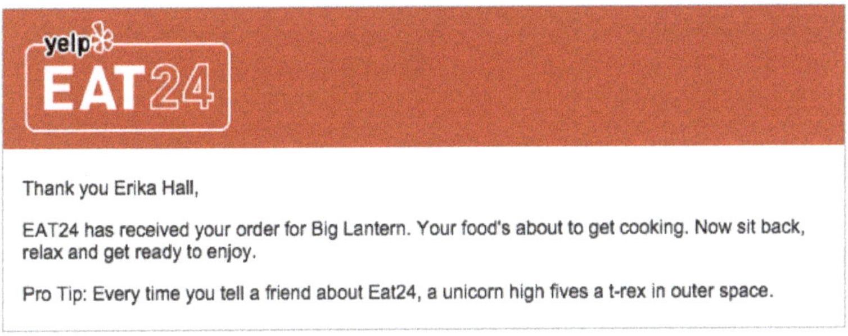

FIG 4.8: Pro Tip: That's not actually a Pro Tip.

Whether and how an interface uses humor is a key part of the tone and personality. Like interaction design, humor depends on context and timing. Playfulness can make a situation more enjoyable, or the attempt can backfire spectacularly. What is funny to one person, might be incomprehensible or offensive to another.

Writing a book leads to ordering a lot of meals online. One of the greatest gifts the internet has bestowed on us is replacing a drawer full of takeout menus and reducing the threat of awkward telephone conversations. In writing this book, I've had the opportunity to review dozens of greetings, reminders, calls to action, "helpful" tips, and status updates about my meals.

After innumerable visits, I've found that the communications start to sound like something out of the Sirius Cybernetics Corporation. Too needy, too chatty, and trying too hard. A unicorn high-fiving a t-rex is mildly funny the first time round (**FIG 4.8**), but by the twentieth time in a month, I want the unicorn to impale the t-rex in the alley behind the Thai place. Interfaces like this arise because designers and writers think of them as abstract descriptors rather than real-life customer interactions—and interactions that happen repeatedly.

Begin with the mood and the state of mind of your customer. In this case, they're likely to be hungry, possibly impatient, and probably also seeking a show on Netflix at the same time.

I asked my buddy Josh for advice on how to avoid unsuccessful attempts at humor:

Here is the easiest way in the world to not be not-funny. Don't try to be funny. The more that you understand the folks you are writing for the more you will find the things that give them pause or tickle their fancy.

Food delivery company Seamless expresses a level of triumphal enthusiasm I save for major accomplishments like finishing marathons or filing my taxes (**FIG 4.9**). They demonstrate how little they know about me with a random list of activities: roof parties, commutes, fight club. Also, they bury the crucial information—how soon is food?—below unnecessary canned cheer.

There's nothing more delightful than being able to summon hot, tasty food to my door with no effort. No need to try so hard. How about:

We estimate your food will be at your door between 8:25–8:35pm. Thanks again for using Seamless and supporting local restaurants!

I'm lucky enough to live walking distance from a small grocery store where I buy most of my food. When I'm not writing a book, I stop in a few times a week. The family who runs it knows me and my shopping habits as well as any online retailer. We usually chat briefly as I check out. Sometimes they comment on my purchases. If one of them ever said, "Here's your receipt! Deliciousness is in the works! Enjoy making dinner while you binge-watch TV or talk to your dog," I would find it odd and uncomfortable.

The British are often very good at navigating tone in verbal humor. *The Guardian* allowed a charmingly exasperated human voice through in a photo caption (**FIG 4.10**). This works because it's truly unexpected. If every photo caption spoke to the reader like this, the effect would wear off quickly.

Some of the best humor comes from playing against expectations. Fast-food brands are generally upbeat about their mass-produced, industrial foodstuff. I'll leave you with one of the darkest voices on Twitter, the parody account Nihilist Arby's (**FIG 4.11**)

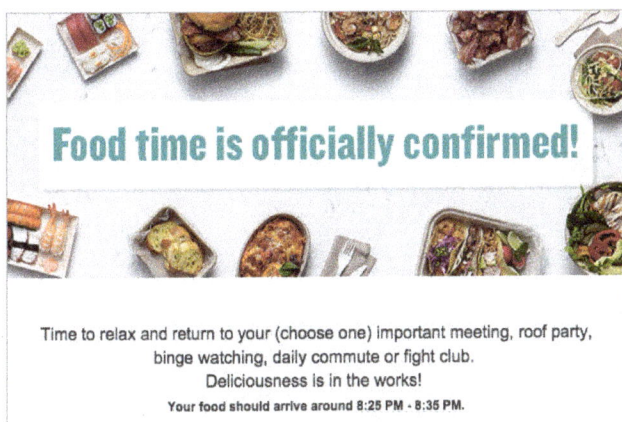

Food time is officially confirmed!

Time to relax and return to your (choose one) important meeting, roof party, binge watching, daily commute or fight club.
Deliciousness is in the works!
Your food should arrive around 8:25 PM - 8:35 PM.

FIG 4.9: Seamless should lead with the important information: when will I eat?!

●●○○○ vodafone UK 📶   14:38   🔋 47% 🔋

theguardian.com   ✕

👤 🔍 •••   theguardian

A cake

📷 Do you know how long I've spent in the Guardian's image archives to find a picture of a bog standard cake? Bloody ages! Thousands of images and not one thing that you'd recognise as a normal cake. This weird ice cream thing is the best I could do. Everything else has a sodding fried egg or a bunch of parsley obscuring it, or has been arranged to resemble a Picasso painting or a sculpture reflecting the futility of existence. When did this place become riddled with hipsters? God, I'm so tired... Photograph: Brian Hagiwara/Getty Images

The triumphant return of the Great British Bake Off shows that everyone in the UK loves cake. So why not put a cake in charge of the Labour party. It might not have the most

FIG 4.10: This is delightful. It's part of a satirical opinion piece on British government (http://bkaprt.com/ cd/04-05/).

Nihilist Arby's
@nihilist_arbys

Follow

Wanna feel old? Very soon you'll be dead and shortly after that no one will remember you ever existed.

Enjoy arbys

1:05 PM - 20 Jul 2017

4,436 Retweets 11,192 Likes

47   4.4K   11K

FIG 4.11: The Twitter parody account Nihilist Arby's plays off the expectation that a major chain's brand voice will be chipper and optimistic.

Expose yourself to excellence

The poet doesn't invent. He listens.
—JEAN COCTEAU

Language is a social phenomenon you are programmed to imitate. Your social self wants to mimic. Borrowing from online banking sites or BuzzFeed will make you parrot the mush without realizing. You'll come out a bureaucrat, or a clickbait zombie.

Strong verbal design is so rare online that looking for best practices risks muddying your thoughts. It's necessary to understand the conversational conventions in existing digital systems, particularly in the domain you're working in, be it shopping or banking or real estate. However, if you want to stand out for your clarity, humanity, and perhaps even for your charm, expose yourself to a wide range of voices in areas of verbal excellence.

Often in my work, I'll scour the internet and app stores for examples of good interface design. And often, I come up short. There always seems to be more clichés and compromises than models to emulate. The same goes for language. It's good to see and hear what exists to identify conventions, but for inspiration, get out into the real (atoms, not bits) world. Go to places where your customers are likely to go. Start listening to how people in different roles *really* talk. Listen for word choice, sentence structure, and tone. Collect and catalog voice samples. Both good and bad. Carry a notebook. Take screenshots. Pop them into a Google doc, print them out and put them on a wall, or drop them into Slack. Describe them and note what makes them effective. Send your team out on Friday to eavesdrop all weekend and report back on Monday.

Also, clean out the clichés and pat phrases by reading some poetry. Aloud. With your team. Poetry is your best source of deliberate intentional language that has nothing to do with your actual work. Reading it will descale your mind, like vinegar in a coffee maker.

I recommend beginning with the Imagists, a group of early-twentieth-century English and American poets. The Imagists favored common speech and novel rhythms to create clear, precise images. (Hence, the name.) Well-known poets include William Carlos Williams, Hilda Doolittle, Amy Lowell, and Marianne Moore.

Ezra Pound, one of Imagism's founders, wrote a manifesto for poetry that would be equally applicable to creating interfaces:

Use no superfluous word, no adjective, which does not reveal something.

Don't use such an expression as "dim lands of peace." It dulls the image. It mixes an abstraction with the concrete. It comes from the writer's not realizing that the natural object is always the adequate symbol.

Go in fear of abstractions. Don't retell in mediocre verse what has already been done in good prose. Don't think any intelligent person is going to be deceived when you try to shirk all the difficulties of the unspeakably difficult art of good prose by chopping your composition into line lengths.

Ezra Pound would probably not be a fan of empty abstractions like "innovative solutions" and the Submit button.

"This is Just to Say" by William Carlos Williams is a terrific example of this compelling economy of language.

I have eaten
the plums
that were in
the icebox

and which
you were probably
saving
for breakfast

Forgive me
they were delicious
so sweet
and so cold

That is a precise image that sticks, capturing an ephemeral moment. No abstractions. All familiar words. It has persisted in popularity since its publication in 1934 and has become the source of many snowclones. (A snowclone plays on the original form, or clones it, but with new words substituted.) The term was coined by linguists Geoffrey K. Pullum and Glen Whitman on the Language Log blog (http://bkaprt.com/cd/04-06/). These poetry jokes are one of the more delightful things to pop up on Twitter with regularity (FIG 4.12).

While reading a bit of verse is unlikely to transform you into the William Carlos Williams of user experience, it will have the same beneficial effect as considering any elegant system or cunning artifact. Strong examples of clear evocative language will inoculate you against creeping banality.

Contemplate the deceptive simplicity, and the way the line breaks hold space in time. Capable and confident writers and designers do not need to embellish. Williams's poem lodges in the brain because it's conversational; the spare voice of the poem is speaking directly to the unseen owner of the plums.

What A Guy
@Yagathai

Follow

Replying to @emilytheslayer

@emilytheslayer This is just to say
I have taken
the kryptonite
that was in
the Batcave
and which
you were probably
saving
to defeat
Clark

7:42 PM - 25 May 2013

stvnrlly
@stvnrlly

Follow

I have closed
the tabs
that were in
the browser

and which
you were probably
saving
to read

Forgive me
they hogged memory
and were
so old

12:06 PM - 30 Jun 2015

FIG 4.12: New variations on this poem seem to appear weekly, such is the power of the form.

Avoidably ugly words

Could I put it more shortly? Have I said anything that is avoidably ugly?
—GEORGE ORWELL, "Politics and the English Language"

And, finally, a fantastic way to clarify voice and values is to discuss avoidably ugly words, those you must never use when speaking for the system you're designing because they are imprecise or empty.

It is often easier to start with the negative space. Think of what you *don't* want to sound like and try that first. Make your interface as bureaucratic or as robotic as you can. Scour Urban Dictionary for verbs you've never heard and slap them on buttons. Then, once you've warmed up and had some fun, give the real voice a try.

Avoidable words might be meaningless filler, taking up space and time without benefitting the user. Or they might be clichés or moldy idioms that any one of a million products or services could use, diluting meaningful differences. I've worked with more than one world-class organization that added "world-class" to the list of words to avoid.

Here's a list of words I wish every organization would avoid for the sake of clarity and interest. This is just a start. I'm sure you'll have a terrific time adding many of your own:

- *Compelling.* This is a word people often use when telling others how to write. It's wishful, not descriptive.
- *Helpful.* Never self-describe as helpful. Let your customers decide that.
- *Quick.* Is it, really?
- *Innovative.* This word used to mean something. Now we've used it up. The same goes for *intelligent* and *smart*.
- *Important.* If you call one thing important, what does that make everything else? Say why it's important, then delete the word *important*.
- *Check out these hot topics.* This phrase adds nothing. You might as well just have a list. Be specific. Describe.

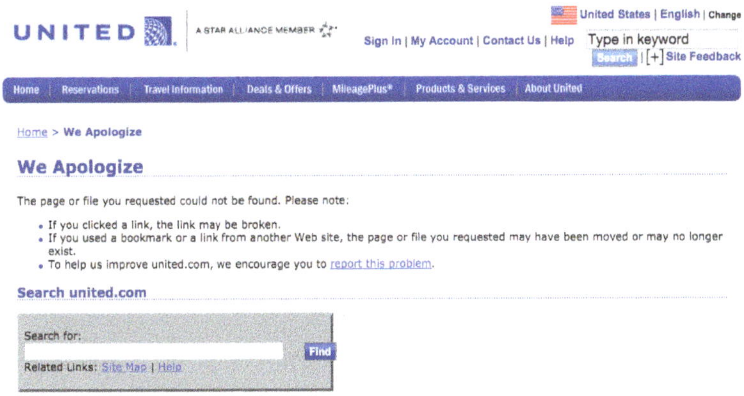

FIG 4.13: My Account is right next to Contact Us. Very strange.

- *Oops!* In 2011, Republican candidate for president Rick Perry said "Oops" when he forgot his answer during a debate and it ended his campaign. Adults restrict the use of this word to inconsequential mishaps, such as spilling a beer on your dog. When you're speaking for a system, choose adult words that show you own your errors.

Remember: If you have to say it, you probably aren't it. And while we're at it, here are a few more functional phrases that are best abandoned in favor of more intentional choices:

- *Submit.* Stop using it as a button label. Forever. Describe the specific action the button triggers.
- *My [items].* As more interfaces speak directly to customers, this usage will eventually disappear. Still, it wouldn't hurt to hurry it along. Otherwise, you'll end up with "We Apologize" and "My Account" colliding, as in United's 404 page (FIG 4.13).
- *Click here.* I've heard arguments for the most generic all to action as an affordance for otherwise obscure links, but it's a crutch for interaction that won't work as interfaces increasingly support switching modes from voice to text and back again.

Making an explicit list of avoidable words is necessary because language is a social phenomenon, and repeating the phrases we hear every day—no matter how threadbare—is the easiest thing to do.

It's all about you

If you've ever spoken to someone who refers to themselves in the third person, or seen an interview with a celebrity who does, it's probably struck you as strange. In natural conversation, people speaking to one another say "I" or "we" and "you."

There's been a lot of back and forth over the use of "my" and "your" in interfaces. Although the roots go deep, let's put this hoary debate down like the zombie attacker it is. Windows 95 introduced My Computer in 1995, followed by My Yahoo!, a customizable dashboard. Then, like mushrooms after a rain, a million mindless imitators emerged. It's as if the user has printed out labels and stuck them to various objects: My Lunch, My Desk, My Red Stapler. Instead of reinforcing a sense of ownership and agency, this feels presumptuous and alienating. For a while the iTunes ioS app navigation bar included both "My Music" and "For You", which is confusing and brought up a lot of feelings about digital rights management (DRM).

To feel conversational, digital systems should be designed to act as true conversational partners. Things belonging to the company that created the system, such as a feature or privacy policy, are "ours." Things belonging to the user, such as a profile or shopping cart, are "yours." Anything that's just a part of the overall experience doesn't necessarily need a possessive pronoun at all.

Amazon has a clear relationship with the people who use their services (**FIG 4.14**). In every instance, they address their customers as "you."

The future may take care of itself in this regard as voice user interfaces (VUI) are becoming more popular, especially for interacting with one's home and car. It's one thing to see an awkward label, and another to hear it spoken aloud: "Now playing David Bowie from my music library." No one wants to

---

Inspired by your browsing history See more

FIG 4.14: This recommendation from Amazon would seem strange and unnatural if the label were "Inspired by my browsing history."

be in a domestic situation in which the digital butler is asserting property rights.

Going non-verbal

Human conversation is uniquely verbal. Among all the creatures on Earth, only we use words. However, the words we use represent only a fraction of what we communicate overall to each other. Because it's possible to incorporate language in a very human way, in translating a human personality to a digital system we lean heavily on language. The same is not true for the nonverbal cues that are a significant part of how humans communicate. Body language, facial expressions, and vocalizations convey essential information that we use to interpret the meaning of the message. A smile, a nod, a touch of the hand, a whisper, or a shout can infuse the same words with warmth, credibility, urgency, or levity.

In the absence of a physical presence, we seek supplemental social cues where we can find them. Over the last few decades, we've learned a lot about computer-mediated communication. Email, texting, and social media generate their own records for us to observe and reflect upon, and these emerging forms of communication are the subject of countless studies, as well as less academic discussions. Most people who communicate online in English recognize that typing in all capital letters is TANTAMOUNT TO SHOUTING. This is a convention.

Digital systems can't use the same body language because, at least for now, computers don't have human bodies. Many of their advantages come from this fact.

A 2015 study by researchers at Binghamton University found that text messages ending with a period seemed less sincere ("Texting insincerely: The role of the period in text messaging" *Computers in Human Behavior* Volume 55, Part B, February 2016). Without explicitly positive information, if you know there's a human on the other side of the computer, it's easy to take a more negative interpretation. Online communities tend towards trolling and flame wars much more than spiraling lovefests.

Computer-mediated interactions, even in a conversational manner, may be less stressful simply because people get less emotionally involved. One of the landmark studies in behavioral economics used magnetic resonance imaging to observe cognitive and emotional processes in people playing the Ultimatum Game with both other humans and computers. In the game, one player proposes a way to divide a sum of money and the other accepts or rejects it. The study showed that participants had a much stronger reaction to perceived unfairness from a computer than from a human. ("The neural basis of economic decision-making in the Ultimatum Game" http://bkaprt. com/cd/04-07/). This study is often cited to explain why people prefer computer-mediated communication, even as compared with humans they know and like.

The challenge of conveying emotion accurately in a few words has led to the popularity of emoticons, emoji, and animated GIFs. The first-recorded emoticon dates to 1635 Slovakia. Ján Ladislaides, the Notary to the Town of Trenčín, drew a smiley next to his signature to indicate the documents he reviewed were in order.

The most appropriate emotional embellishment depends on the context of the communication and the personality of the system. I've already mentioned the Quartz news app which includes emoji as part of its interface, but it would probably be off-putting for your bank to text you a sad face to let you know that your account is overdrawn. For many systems, creating appropriate nonverbal cues requires specialized visual, industrial, and sound designers. For any of these to be effective, however, they require a personality that's clearly defined, and articulated in words.

EXCELSIOR!

It all comes down to valuing your customer and knowing your values. Tacking on a friendly face in bad faith won't make the underlying system more meaningful or sustainable. That's a tough act to keep up.

Products and services that succeed in having the right amount of the right personality do so because thoughtful humans cared enough to be honest and intentional. Clear, collaborative conversations are important throughout the entire process. Next, we'll look at how to work with your team to do this. Creating a design process is just another type of interaction design. Designers are also people with habits and biases who work in a context.

Easy-peasy, right?

5 GETTING IT DONE

> " *Conversation unites a team around a shared vision. It also brings insights from different disciplines to the project much earlier than a traditional design cycle would allow. As new ideas are formed or changes are made to the design, a team member's insight can quickly challenge those ideas in a way the designer alone might not have recognized.*"
> —JEFF GOTHELF, *Lean UX*

TO CREATE MORE CONVERSATIONAL, human-centered systems, we need to work in a more conversational, human-centered way. This is a challenge because the way we do business is still largely driven by documentation and hierarchy. Doing business requires a certain amount of both. Innovating in the space between stasis and anarchy requires a delicate balance.

The working style of an organization determines what it's capable of creating. In 1967, the computer scientist Melvin Conway published a paper titled "How Do Committees Invent?" The central thesis came to be known as Conway's Law in the software community. Dr. Conway summarizes it on his website:

Any organization that designs a system (defined broadly) will produce a design whose structure is a copy of the organization's communication structure.

The first design proposed is very rarely the final iteration. A design project is a series of decisions. How an organization makes decisions will determine the extent to which the skills of each individual contribute to the whole. It's a web design cliché that a company's first website tends to mirror its org chart rather than the needs of its customers. Even designs that aren't so directly representative will reflect the processes and priorities from which it emerged.

A half century after Conway's observation, more fluid organizations are producing, or contributing to, increasingly dynamic and complex systems. These systems are connected by contextual associations and interactions over time rather than by a stable structure. I propose a corollary:

The degree to which a system feels human and goal-oriented in its interactions reflects how well its creators interacted with each other.

A harmonious interface is the product of functional, interdisciplinary communication and clear, well-informed decision-making. Systems with visible seams are the result of hand-offs and unresolved arguments.

It's clear when the legal team has taken ownership of a step in the sign-up process and everything grinds to a halt in a wall of text. An interface of overwhelming choices represents territory battles or general aimlessness. Indecipherable error messages indicate that the design team wrapped up and went on vacation right about the time the engineering team took over. Or that these essential parts of the interaction weren't considered from the beginning.

Design is a machine for making new things in existing contexts, and for making old contexts obsolete. Instead of machining new factory tools, we fabricate processes and types of artifacts. The fundamental paradox of design is that the process by which we bring new things into the world can itself become a comfortable barrier to change. We work in the ways we know how to work. We need to balance the efficiency of mastery against the desire for novelty and the need for new solutions.

An organization is a set of people with shared purpose and processes—that is to say, shared habits. Changing habits is hard. In the absence of a crisis, it requires tremendous effort to change a habit, even if the current way of doing things causes pain and the new way will be easier and more effective. "Process" is just a business word for habit, or an aspiration to develop that habit.

Creating a design process is just another type of interaction design. A group of people sharing a system need to behave a certain way and have access to the right information at the right time to get to a successful outcome.

WHAT NEEDS TO CHANGE

We recently worked with a very successful software company that was committed to design and working in a user-centered, multidisciplinary way. However, even though each product had a dedicated visual designer and a dedicated interaction designer, writing was treated as a separate category of work—just as with the publishing company I mentioned at the very beginning of this book.

One writer was assigned to five or ten products at a time. And even though the writers were contributing to software, as the visual designer and interaction designer were, their work was treated in a much traditional way: it went through a typical editing workflow of review cycles where sign-off was undervalued and second-guessed by people in charge. Even after rounds of editing and revision, anyone with sufficient power had the ability to make changes by fiat.

And, most damaging of all, there was no feedback loop to let the writers know to what extent their work had been successful, which means they could never learn and improve. They continued to rely on their expertise to manage their contribution to this highly fragmented process.

For their part, the writers were protective of their work. Several of them spoke of their "creative process," which they perceived as having to take place in reflective isolation. They held fast to the creation of a first draft sufficiently polished to submit for feedback.

This was a process that had served the creation of documents, but did not make sense for interactive systems. Splitting the design in this way—with verbal components running on a separate track—did a disservice to the entire experience. The team was multidisciplinary in the sense that people with different skill sets contributed to the work, but it wasn't collaborative. This lead to a user experience of boxes filled in with language. Authority ruled the day, rather than the inspiration of wit or finding the fitting phrase for a particular moment.

Identifying the problem

Process change begins with a problem. If there's no perceived problem, there will be no change. A more conversational design process will help solve a variety of problems that fall into two categories, wasting resources, and missing opportunities (which is just another kind of waste).

The words and pictures divide

Interaction and interface design must include decisions about verbal language and nonverbal cues as part of the same process and at the same time. Since teams are often divided by their tools, and the artifacts those tools generate, this is the greatest challenge. There are few tools that support cross-disciplinary collaboration.

A lack of innovation

The greatest barrier to innovation is comfort with the current way of doing things. A process based in documents or artifacts makes it easy to keep repackaging familiar thinking in newer, shinier ways and still feel like you're making progress. Stepping away from what you already know brings opportunities into view.

Confusing polish for value

Teams that aren't comfortable communicating across traditional discipline divisions will have a strong urge to create a thing to demonstrate their value in the design process. Once an artifact exists, it's easy to invest it with value and be unwilling to discard the weak idea it may represent.

A systems-oriented view

Agile and Lean approaches, at their core, offer ways to build software better, not solve problems for people and business. These processes risk conflating features with value—the faster the team builds and the more they build, doesn't equate with the value they're producing. The goal of conversational design, and any good design, is to create the most value with the fewest features. Just like it's better to get the message across in the fewest words, and solve the problem in the simplest way. More thinking, less building.

The Tower of Babel

People cling to the language of their disciplines, business, engineering, writing, design. Humans love to do the in-group/out-group thing. Get people to talk together early and it makes everything clearer down the line. An organization can't solve a problem holistically if different groups are working from different information sets. Talking together is often a more effective way to share information than passing documentation around.

THE KEYS TO COLLABORATION

Being conversational in your approach to design work doesn't mean sitting around chatting over coffee, although that's a fine thing to do. It means that the work culture reflects collaborative social values that invite participation and support immediacy, rather than a highly-political, hierarchical, document-based way. Like the digital interactions we discussed in Chapter 2, each value of conversational culture has an analog in practice:

- **Cooperative**. The products of our culture reflect and influence one another. Team members support each other and work towards mutual success.
- **Goal-oriented.** All efforts are in support of a clear goal.
- **Context-aware.** Priorities, constraints, and real-world conditions are brought to bear in all problem-solving towards the goal.
- **Quick and clear**. Everyone does their part to keep things moving along and works together to eliminate fuzzy thinking.
- **Turn-based.** Smart, highly-verbal professionals are trained and rewarded to be good talkers. Despite the vogue for active listening a few years ago, listening is an underrepresented skill in the workplace. For this whole conversational thing to work, everyone must wait their turn and truly listen to the person speaking. All participants have a chance to contribute and respond to the contributions of others. It's understood that the work is an interactive, collective process rather than the sum of individual contributions.
- **Truthful**. Candid assessments and complete information are welcome and expected.
- **Polite**. There's a shared set of communication norms that are respected.
- **Error-tolerant**. Everyone works to create a safe environment in which people can make suggestions and offer ideas that may be wrong or occasionally fail, and this is seen as a productive part of the process.

These are the qualities of a work culture that supports and recognizes the humanity of every member of the team. And this style of interaction will enhance any team's ability to create systems that feel human, humane, and lively.

Organizations of any scale or complexity require a certain amount of authority and documentation to function. This is a fact rather than a flaw. However, all organizations should have clear goals and a willingness to continuously reflect and improve.

FORM FOLLOWS MEANING

Don't believe the notion that IxD is over & we can just use patterns. Most software we use is still crappy because of concepts, not buttons.

—ALAN COOPER via Twitter

Once you have the people in your organization working in a more immediate and conversational manner, you can introduce other process revisions. The big one is rethinking each phase of the interaction and interface design process with the aim of eliminating the reliance on specific artifacts or documentation as a proxy for design and understanding.

For people raised on screens, or hooked on screens, it's tempting to start drawing screens. As soon as you start drawing screens, you stop being truly human-centered.

Rather than starting with sketches or blocking out shapes, try the Concept-Script-Sketch model (FIG 5.1) to ensure that your ideas and the pace of action are on target.

Every design needs to start with a strong concept, the big idea, the reason to exist. It doesn't matter how delightful the surface styling is if the underlying idea is weak. And a strong idea can be device-independent or exist across multiple devices or modes of interaction.

Next, move onto the script, the exchange of information that breathes life into the interaction. This is the core of the interaction. Try working collaboratively all together in front of a whiteboard, or using whatever communication platforms

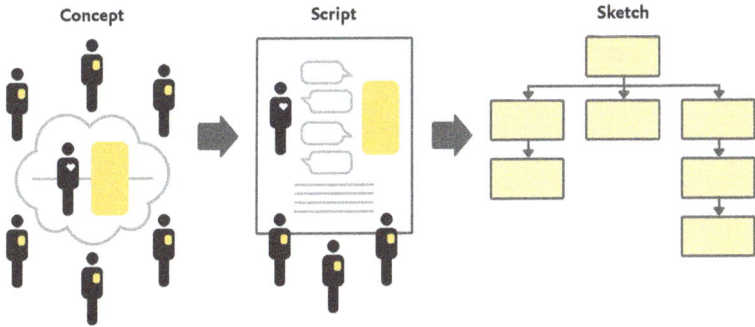

FIG 5.1: It may feel unnatural to postpone visualizing the interaction spatially until after conceptualizing the interaction, and scripting the exchange of meaning in time. Conversation offers a powerful way to take a truly human-centered approach to designing.

feel most natural for your team. Something as simple as a text editor might be a good place to save ideas. Or you could try an interactive story editor like Twine. Taking a multimodal, multidevice approach to interaction design is a new and complex practice, so there aren't specific tools for it.

Finally, sketch the interface of the system that supports that exchange of information or meaning. Show how it fits into the larger context of the user's life.

This process will allow you to consider the same exchange of value through multiple channels and interfaces.

Give it your minimum

A core concept of Lean UX is the Minimum Viable Product (MVP). But I suggest stepping back even further. MVP still centers the system—it can be a fine technique for iterating on solutions—but only after spending more time with the design questions. Design by prototyping and testing is like answering a question before checking to see whether anyone had asked. MVP can be great for working within existing systems, but what if you want to solve a higher-order problem? What if you don't even know if something is a product yet? It's not about testing ideas, rather it's about understanding the exchange of value. Step away from features to a higher level, it's the thought experiment before the experiment. Otherwise it's easy to keep patching bad ideas with features.

I propose the Minimum Meaningful Conversation, a lightweight way to think outside the system and see the value you purport to offer from your customer's perspective. Rather than think of the product as a series of screens in space, consider the moments in time.

Start from the customer's intent, the questions they ask, the needs they express. How do you answer, and where do you answer? Too often this moment is left up to marketing. How you get in front of your customer at the moment they have a problem is a core part of the design. If you don't figure that out, it doesn't matter how good your proposed solution is.

Some key questions to ask:

- How will the customer express their need? How will you understand their intention?
- What exchange happens at that point?
- What does the customer have to give you in order for you to solve their problem?
- What choices do you give them?
- How do you close the conversation impressing the customer that you gave them something of value?
- How do you leave the door open to future interactions?

To think through the value exchange in more detail, use the Minimum Viable Conversation Worksheet (FIG 5.2).

Prototyping and testing

The goal of early prototypes is to separate the polish of the production from the value of the idea. Because, as I've mentioned, it's easy to fall in love with and feel defensive about labor-intensive artifacts, even if they're merely shiny vessels for weak ideas.

The best way to design interactive systems is to interact with people. And that's the best way to prototype and test them, too. It takes a lot of work to create prototypes for a wide variety

Minimum Viable Conversation Worksheet

	CUSTOMER	SYSTEM
Context	What's happening? Where are they? What tools, abilities, and information do they have with them?	Where's the system's presence in this context?
Event	What happens to make the customer aware of the problem?	How does the system become aware of the customer?
Intention	What's the customer's response to the problem? How do they express their need?	How does the system register this intention?
Introduction	How does the customer need to be identified to the system?	How does the system make itself appear meaningful and credible?
Orientation	What's the customer's model of the conceptual space?	How does the system establish and communicate the boundaries?
Action	What motivates the customer to interact with the system?	What action is offered? What's the value of that action to the customer?
Guidance	What assistance might the customer need to complete the action successfully?	How does the system support the success of the customer?
Error	What could go wrong?	How does the system help the user get back on track?
Closure	How does the customer know the interaction has concluded successfully?	How does the system finish strong and plant the seeds for further interaction in the future?

FIG 5.2: Your team should be able to work through this together before creating any sketches of solutions.

of flows and specific language. You can test more interactions much faster by doing less. Just pretend to be the ghost in the machine and note what works as you go.

In literature, the "Wizard of Oz" refers to an ordinary man who controls machines from behind a curtain so that he can pretend to be a magical, floating head on a throne. In the Wizard of Oz methodology, a human "wizard" simulates the behavior of an interactive system by intercepting all communications between the user and the system. As a child of the 1980s, I think a more apt analogy is Max Headroom. Max Headroom originated as the wise-cracking, "computer-generated" host of a music video show (FIG 5.3). Since computer technology was not yet up to the task of rendering an interactive character of any sophistication, the computer was played with glitchy verisimilitude by an actor encased in fiberglass and filmed against a blue screen. (The makeup artist responsible for the four-hour transformation was a bit peeved that the show—in order not to tip their hand—never competed for any awards in his category.)

To prototype and test flow and conversation interactions, a human can perform the tasks of a system—processing user input and providing output. Messaging and voice applications might seem the most natural things to test this way, but you can also use the technique to test screen-based interfaces using sketches. The human proxy can be totally obvious, like a visible puppeteer, typing words that appear on the user's screen or saying the words that would come out of a speaker attributed to machine intelligence. You can let your test participants know that they're interacting with a human, or, for the hidden homunculus, tell them that it's a programmed digital system. If the participant is a good representative of your target customer, they'll be able to suspend disbelief, even if the human is present. This practice removes the limitation of designing based on technological constraints, particularly in terms of interaction logic. You can glean a good sense of customer expectation this way.

One you have a better sense of what the ideal interaction is from the user's perspective, it may take some creative problem solving to meet expectations with available design and technology resources. As Max Headroom showed, it's often possible to

FIG 5.3: Max Headroom was a "computer-generated" character from the mid-1980s. He was played by an actual human under four hours of makeup and a fiberglass tux. Lucky for you, pretending to be a digital interface has gotten much easier.

offer the sensation of greater machine intelligence by applying human ingenuity.

This type of prototyping calls on a very different set of skills to those typically associated with design. It can be uncomfortable for someone used to designing screens to play the part of the system itself. Going a step further and stepping into the role of the system not only saves time and effort in designing and coding, but it can truly test the core of the interaction from both sides.

THE BEST SMALL CHANGES

All work, even design work, is a practice grounded in a set of habits. And, as mentioned before, changing habits is hard, especially changing them at scale. Anything you can do to work in a more goal-driven, interdisciplinary, and conversational manner, even if only in a small way, is a good thing.

- **Reiterate the goals and principles.** This may feel redundant, but forgetting is easy and repetition is powerful. Surgeons and pilots use checklists to make sure they don't kill anyone. You can use checklists to ensure that you aren't killing time.
- **Work in real time.** Resolve to find ways to include all perspectives at the same time rather than reverting to passing an artifact around for commentary. To reinforce the sense of a team working together on the same objective, incorporate real-time collaboration tools such as Slack into your process and eliminate internal email as much as possible.
- **Encourage candid feedback.** Everyone must practice giving, receiving, and responding to feedback. Feedback and critique are the essence of a lively, iterative process.
- **Include decision-makers.** They might balk at adding another meeting to their schedule, but it will save so much time down the line, even if their participation is remote.
- **Talk about decisions over artifacts.** Frame the design conversation around creating an experience and exchanging information in time, rather than laying out elements in space.
- **Never permit lorem ipsum.** No placeholder language. Language and meaning belong at the center of the experience. Reinforce the idea that specific language is subject to continuous iteration and is a part of the design process. It's not some parallel but different "writing" process.
- **Read aloud every word** that's intended to be part of the customer-facing design. This is an essential test for meaning and timing, and the only way to ensure that your interface works across modes in both text and speech. It will also lighten the weight and perceived permanence of anything seen as "the written word."

BEING HUMAN TOGETHER

It was as if my whole life revolved around trying to judge the appropriate point in a conversation to say goodbye.
—HARUKI MURAKAMI, *Blind Willow, Sleeping Woman*

It may seem daunting to change how your organization, or your client's organization, approaches design and collaboration. Design is a practice. And significant change is something that happens incrementally over time with everyone involved contributing to the practice.

People are complicated and messy. All too often, digital tools and processes are used to avoid dealing directly with one another—even while designing tools and processes for other people. The true purpose of process is to make humans more efficient, not eradicate the humanity. When you look at a problem you want to solve with design, look deep. Look past the surface to what it means—not to the components or code, but to the human beings.

Different approaches will work for different teams. The better you can cultivate direct, honest communication and respect, the more easily you'll be able to collaborate to create meaningful interactions.

It's exciting stuff working with interactive, interconnected design. Every year brings something that used to be the stuff of science fiction. I'm certain that in the near future, we'll be able to control computers with our minds. Telepathy and telekinesis will become practical realities. And even these abilities won't change anything fundamental about human nature.

If you want to invent the future, you would do well to review the last several hundred thousand years. Humans have always been the same creative, social, insecure, argumentative creatures looking for meaning. More of us wear pants these days, it's true. But communicating with one another is our greatest joy—and terror. Using machines as intermediaries allows us to feel close, but not too close. And it can make things weird. Every day is a new opportunity to avoid making things weirder than they need to be.

This is what your work can offer. And this is how you can work with others to make it happen. Know your goal. Listen before you speak. And create a system that sings.

Often the best way to start a conversation is with a question: What are we here to do?

ACKNOWLEDGEMENTS

Endeavoring to write one book may be regarded as naïveté. To write a second is just ridiculous. This book would not have been possible without the unwavering faith, advice, and encouragement emanating from the bright shining light that is Katel LeDû. Gratitude is due as well to the steadfast Jeffrey Zeldman and Jason Santa Maria, who are truly gentlemen apart. Words cannot express what I owe to the exacting fortitude, compassion, and wit of my editors, Lisa Maria Martin, Kate Towsey, and Sue Apfelbaum, whom I would gladly endorse on LinkedIn for remote psychotherapy. Jon Long created the delightful illustrations from some rather tragic diagrams.

The ideas encapsulated here evolved from wisdom accrued through many hours of more or less enthusiastically shouty exchanges in every possible channel with Kristina Halvorson, Joshua A. Cagan, Ben Brown, Ross Floate, Austin Kleon, Jesse Taggert, Arjun Basu, and Anna Picard. Stewart Scott-Curran's invitation to speak caused some key ideas to coalesce. Larisa Berger was a steady source of both ideas and encouragement. Kio Stark stayed by my side in cyberspace. Kayla Cagan is a first-rate, no-drama dramaturge, writer, poet, and all-around inspiration. Amy Jo Kim kept me energized. And I found that Bill deRouchey and I share strong opinions about buttons.

Huge gratitude to the big-hearted lovers of language Erin McKean, Margo Stern, Karen Wickre, Shauna Wright, and Sam Soma. Long ago, Jeff Tidwell admonished me to "ride the wave" and I've striven ever since to follow his advice.

It turns out that developers often have the best way with words. David McCreath, Steph Monette, Jim Ray, Liam Campbell, and Miguel Ceja are fluent in lyrics, polemics, and jokes, as well as code. Mike Essl, Tom Carmony, Amber Costley, Andy Davies, and John Hanawalt demonstrate that visual designers are tremendous conversationalists. And Sonia Harris is always there with the perfect swear.

Over the years, I've had the great fortune to work with clients who made me a better designer and communicator through the benefit of their collaboration, including Heather Walls, Zack McCune, Melody Kramer, Katherine Maher, Ashley Budd, Thomas Deneuville, Susan McConkey, Jarrod Bell, Mark Jannot, John Mahoney, Kevin Cosgrove, John Hassell, Chipp Winston, Laura Silber, Scott Stocker, Alex Bloomingdale, Brad Satoris, Chris Westfall, and Albert McMurry. So many conversations, and not a few drinks.

In everything I do, I am just following in the footsteps of my design heroes and colleagues Jared Spool, Dana Chisnell, Alan Cooper, Laura Klein, Khoi Vinh, Dan Brown, and Louis Rosenfeld. John Maeda is the soul of graciousness. Jennifer Daniel, the emoji GOAT, is a force of nature and style. Robyn Kanner just plain kicks ass. There are so many more people to mention, I could fill another hundred pages.

Henry Monteiro always inspires me with his fearless approach to writing. And, of course, this wouldn't have been possible without the support of Mike Monteiro, who stayed positive even when I was at my most annoying. He made many dinners for me, saw a whole lot of movies without me, and always made sure Rupert had a walk. (Mike is going to say I made this about the dog.)

RESOURCES

To solve new problems in new ways, it helps to look beyond the typical design reading list and get a sense of historical perspective. Here are a few books and articles that address the nature and importance of language as well as what it means to create useful systems for humans.

Know your history

To create the future, it's necessary to understand the past.

- *Orality and Literacy*, Walter J. Ong.
- "The Gutenberg Parenthesis," Thomas Pettit. Pettit argues that the invention of the printing press was simply an interruption to conversational culture (http://bkaprt.com/cd/06-01/).
- "Politics and the English Language," George Orwell. No one knew better about the uses and misuses of the power of language (http://bkaprt.com/cd/06-02/).

Think about thinking

- *The Stuff of Thought*, Steven Pinker. Dense, entertaining read about language and human nature. The part about profanity is illuminating.
- *Metaphors We Live By,* George Lakoff. The classic book explains how metaphors shape cognition and behavior in every arena of human life.
- *Behave: The Biology of Humans at Our Best and Worst*, Robert Sapolsky. This is some real, first-principles stuff. Everyone who designs for people needs to read it.

Apply it to systems

There are plenty of books out there with tips and techniques, but these go deeper, and in an accessible way.

- *Language and Communication*, Agnes Kukulska-Hulme. This is the best analysis I've found of using language as part of computer-system interfaces. Out of print. Find it.
- *Reclaiming Conversation: The Power of Talk in a Digital Age*, Sherry Turkle. Based on five years of research, this book argues for the importance of talking to actual other humans.
- On Poetry, Glyn Maxwell. These thoughts about time and the relationship between meaning and empty space are essential for all designers.
- Design for Voice Interfaces, Laura Klein. A clear, brief report that outlines the core considerations.
- Interview with Anna Pickard of Slack (http://bkaprt.com/cd/06-03/).
- Intercom, a customer messaging company with an unusually thoughtful blog about conversational design (http://bkaprt.com/cd/06-04/).

Create an exchange of value

This work is in service of commerce and society, after all.

- *How to Win Friends and Influence People*, Dale Carnegie. I often call this the only interaction design book everyone must read. The stories are adorably old-fashioned. The principles are strong.
- *Implementing Value Pricing: A Radical Business Model for Professional Firms*, Ronald J. Baker. The cover "says" dry business textbook, but it's much more of an entertaining meditation on creating a sustainable exchange of value.

REFERENCES

Shortened URLs are numbered sequentially; the related long URLs are listed below for reference.

Chapter 1

01-01 http://ideas.time.com/2013/04/25/is-texting-killing-the-english-language/
01-02 http://www.pewinternet.org/2015/04/01/us-smartphone-use-in-2015/
01-03 https://www.ncbi.nlm.nih.gov/pubmed/24023634

Chapter 2

02-01 https://www.nngroup.com/articles/get-started/

Chapter 3

03-01 https://www.experian.com/innovation/thought-leadership/amazon-echo-consumer-survey.jsp
03-02 https://material.io/guidelines/patterns/notifications.html
03-03 http://berglondon.com/blog/2010/09/04/b-a-s-a-a-p/

Chapter 4

04-01 https://www.contagious.com/blogs/news-and-views/19490116-slacks-editorial-soul-anna-pickard-on-writing-the-brand-experience
04-02 https://www.fastcodesign.com/3054934/the-next-phase-of-ux-designing-chatbot-personalities
04-03 https://medium.com/startup-grind/google-home-vs-alexa-56e26f69ac77
04-04 https://developer.amazon.com/blogs/alexa/post/ee2b4898-8612-4a11-8c3f-10bd13837442/ssml-speechcons-in-alexa-skills-now-available-in-the-uk-and-germany
04-05 https://www.theguardian.com/science/brain-flapping/2015/aug/06/jericho-the-lion-alternative-labour-leadership
04-06 http://itre.cis.upenn.edu/~myl/languagelog/archives/000350.html
04-07 https://www.ncbi.nlm.nih.gov/pubmed/12805551

Resources

06-01   https://commforum.mit.edu/the-gutenberg-parenthesis-oral-tradi-
        tion-and-digital-technologies-29e1a4fde271

06-02   http://www.orwell.ru/library/essays/politics/english/e_polit

06-03   https://www.contagious.com/blogs/news-and-views/19490116-slacks-ed-
        itorial-soul-anna-pickard-on-writing-the-brand-experience

06-04   https://blog.intercom.com/category/design/

INDEX

ABOUT A BOOK APART

We cover the emerging and essential topics in web design and development with style, clarity, and above all, brevity—because working designer-developers can't afford to waste time.

COLOPHON

The text is set in FF Yoga and its companion, FF Yoga Sans, both by Xavier Dupré. Headlines and cover are set in Titling Gothic by David Berlow.

This book was printed in the United States using FSC certified papers.

FSC
www.fsc.org

Erika Hall has been working in web design and development since the late twentieth century. In 2001, she cofounded Mule Design Studio, where she leads the strategy consulting practice. Her enthusiasm for evidence-based decision-making led her to write *Just Enough Research*. She speaks frequently to international audiences on topics ranging from collaboration and design research to effective interface language. Her current talks explore the limits of using quantitative data to make design decisions.